# SEARCH GAMES

Academic Press Rapid Manuscript Reproduction

This is Volume 149 in
MATHEMATICS IN SCIENCE AND ENGINEERING
A Series of Monographs and Textbooks
Edited by RICHARD BELLMAN, *University of Southern California*

The complete listing of books in this series is available from the Publisher upon request.

# SEARCH GAMES

*Shmuel Gal*
IBM Israel Scientific Center
Haifa, Israel

**ACADEMIC PRESS**
*A Subsidiary of Harcourt Brace Jovanovich, Publishers*
**New York London Toronto Sydney San Francisco 1980**

ACADEMIC PRESS, INC.
111 Fifth Avenue, New York, New York 10003

*United Kingdom Edition published by*
ACADEMIC PRESS, INC. (LONDON) LTD.
24/28 Oval Road, London NW1 7DX

LIBRARY OF CONGRESS CATALOG CARD NUMBER: 80-68856

ISBN 0-12-273850-0

PRINTED IN THE UNITED STATES OF AMERICA

80 81 82 83 9 8 7 6 5 4 3 2 1

*To my parents*
*Haya and Nachman*

# CONTENTS

# PREFACE

This monograph deals with the problem of finding optimal search trajectories in order to locate a target. The general approach used is to look upon the situation as a game between a searcher and a hider and to present optimal strategies for both participants in this game. In some of the problems, the hider will be assumed to be stationary, while in others, the hider is allowed to move and evade the searcher as long as possible. In most cases, we shall assume that there is one hider and one searcher, but sometimes we shall show how the results can be generalized to include the case of several searchers and one hider.

This monograph is written mainly for people who are interested in search and minimax problems. In order to read and understand the book, one needs only a basic knowledge of game theory and probability. (Actually, the basic notions of game theory needed are described in Appendix 4.) Going through all the details of some of the proofs, however, requires a slightly deeper mathematical understanding. (Some of the details of the proofs can be skipped on the first reading.) Some interesting open problems are presented, and thus it is hoped that those who are interested in doing further research in this field will find stimulating material in this monograph.

Guide to Dependence of Chapters

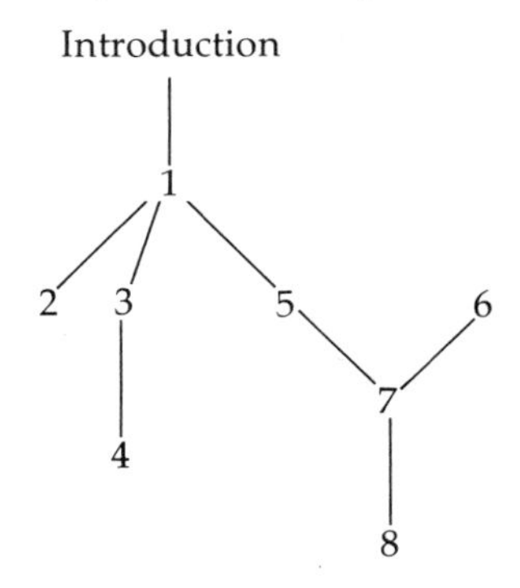

# ACKNOWLEDGMENTS

I wish to express appreciation to the staff members of the IBM Israel Scientific Center. I am especially grateful to Vardy Amdursky and Shmuel Katz for reading the manuscript and for their helpful comments, to Dan Chazan for reading and commenting on Appendix 1, and to Abraham Ziv for providing me with Lemma 2 of Section 4.5.

I appreciate the valuable comments made by Nimrod Megiddo of Tel Aviv University and by Eliahu Shamir of the Hebrew University, who read the manuscript. Several helpful discussions were also held with Donald J. Newman of Temple University.

I wish to thank Raya Anavi for typing a part of the draft.

Above all, I would like to express my deep appreciation to my wife Liza and to my children Ofer and Vered for their patience during the long hours which I spent working on this book. My wife Liza also helped in typing a part of the draft, and was a source of permanent encouragement to me.

# Frequently Used Notations

| | |
|---|---|
| [ ] | Integer part |
| $\mu$ | Lebesgue measure of the search space |
| $\mu^*$ | Minimal length of a tour which covers the search space |
| $\nu(Z)$ | Maximal velocity of the searcher |
| $C(S, H)$ | Cost function (the payoff to the hider) |
| $c(s, h)$ | Expected cost |
| $\tilde{C}(S, H)$ | Normalized cost function |
| $\tilde{c}(s, h)$ | Expected normalized cost |
| $d(Z_1, Z_2)$ | Distance between $Z_1$ and $Z_2$ |
| D | Diameter of the search space |
| E | Expectation |
| g | Rate of discovery of the searcher |
| H | A pure hiding strategy |
| $\|H\|$ | Distance of an immobile hider from the origin |
| h | A mixed hiding strategy |
| $h_R$ | Completely randomized hiding strategy |
| iff | If and only if |
| O | Origin (usually, the starting point of the searcher) |
| Pr(A) | Probability of an event A |
| Q | Search space |

| | |
|---|---|
| $r$, $r(Z)$ | Discovery radius |
| $S$ | A pure search strategy (a search trajectory) |
| $s$ | A mixed search strategy |
| $T$ | Capture time |
| TH(Th) | Set of all pure (mixed) hiding strategies |
| TS(Ts) | Set of all pure (mixed) search strategies |
| $t$ | Time parameter |
| $v(H)$, $v(h)$ | Value of the hiding strategy ($v(h) = \inf_s c(s, h)$) |
| $v(S)$, $v(s)$ | Value of the search strategy ($v(s) = \sup_h c(s, h)$) |
| $\overline{VP}$ | Minimal value obtained by a pure search strategy (the "pure value") ($\overline{VP} = \inf_S v(S)$) |
| $v$ | Value of the search game ($v = \inf_s v(s) = \sup_h v(h)$) |
| $w$ | Maximal velocity of the hider |
| $Z$ | A point in the search space |

# Introduction

In the search problems considered in this monograph, it is necessary to detect an object such as a person, a vehicle, or a bomb. An extensive research effort, initiated by Koopman and his colleagues (1946), has been carried out on this type of problem. Most of the previous works, however, assume that there exists a probability distribution for the location of the target, which is known to the searcher (i.e., a Bayesian approach is used). In addition, most previous works, such as those considered in "Theory of Optimal Search" (Stone, 1975), are concerned with finding the optimal distribution of effort spent in the search but do not actually present optimal search trajectories. This fact has already been pointed out by Dobbie (1968) in his survey on search theory.

In this monograph we shall take into consideration the fact that usually the searcher has to move along a continuous trajectory and attempt to find those trajectories which are optimal in the sense described in the sequel. We shall not assume any knowledge about the probability distribution of the hider's location, using instead a minimax approach. The minimax approach can be interpreted in two ways. One could either decide that because of the lack of knowledge about the distribution of the hider, the searcher would like to assure himself against the worst possible case; or, as in many military situations, the hider is an opponent who wishes to evade the searcher as long as possible. This approach leads us to view the situation as a game between the searcher and the hider.

In general, we shall consider search games of the following type. The search takes place in a set $Q$ to be called "the search space." We shall distinguish between games in compact search spaces which are considered in Part I and games in unbounded domains which are considered in Part II. The searcher usually starts moving from a specified point $O$ called the origin and is free to choose any continuous trajectory inside $Q$, subject to a maximal velocity constraint. As to the hider, in some of the problems it will be assumed that the hider is immobile and can only choose his hiding point, but we shall also consider games with a mobile hider who can choose any continuous trajectory inside $Q$. It will always be assumed that neither the searcher nor the hider has any knowledge about the movement of the other player until their distance apart is less than or equal to the discovery radius $r$, and at this very moment capture occurs.

Each search problem will be presented as a two-person zero-sum game. (The basic notions of such games are described in Appendix 4.) In order to treat a game mathematically, one must first present the set of strategies available to each of the players. These strategies will be called "pure strategies" in order to distinguish between them and probabilistic choices among them which will be called "mixed strategies." We shall denote the set of pure strategies of the searcher by

TS and the set of pure strategies of the hider by TH. Any $S \in TS$ is a continuous trajectory inside Q such that S(t) represents the point which is visited by the searcher at time t. As to the hider, we have to distinguish between two cases: If the hider is immobile, then he can only choose his hiding point H. If he is mobile, then his strategy H is a continuous trajectory H(t) so that, for any $t \geq 0$, H(t) is the point occupied by the hider at time t.

The next step in describing the search game is to present a cost function (the payoff) C(S, H), where S is a pure search strategy and H is a pure hiding strategy. The cost C(S, H) has to represent the loss of the searcher (or the effort spent in searching) if the searcher uses strategy S and the hider uses strategy H. Since the game is assumed to be zero-sum, C(S, H) also represents the gain of the hider, so that the players have opposite goals: The searcher wishes to make the cost as small as possible, while the hider wishes to make it large. The natural choice for the cost function is the time spent until the hider is captured. For the case of a bounded search space Q, this choice presents no problems, but if Q is unbounded and if no restrictions are imposed on the hider, then he can make the capture time as large as desired by choosing points which are very far from the origin. We overcome that difficulty either by imposing a restriction on the expected distance of the hiding point from the origin or by normalizing the cost function. The details concerning the choice of a cost function for unbounded search spaces are presented in Chapter 5.

Given the available pure strategies and the cost function C(S, H), the value v(S) guaranteed by a pure search strategy S is defined as the maximal cost which could be paid by the searcher if he uses the strategy S; thus,

$$v(S) = \sup_{H \in TH} C(S, H). \tag{1}$$

Let

$$\overline{VP} = \inf_{S \in TS} v(S). \tag{2}$$

Then for any $\varepsilon > 0$, the searcher can find a pure strategy which guarantees that the loss will not exceed $(1 + \varepsilon)\overline{VP}$. A pure strategy $S_\varepsilon$ which satisfies

$$v(S_\varepsilon) < (1 + \varepsilon)\overline{VP} \tag{3}$$

will be called an "$\varepsilon$-minimax search trajectory." If there exists a pure strategy S which satisfies

$$v(S) = \overline{VP}, \tag{4}$$

then S will be called a "minimax search trajectory." $\overline{VP}$ is called the "pure value of the game."

The value $\overline{VP}$ represents the minimal capture time which can be guaranteed by the searcher if he uses a fixed trajectory, but in all the interesting search games the searcher can do better on the average if he uses random choices out of his pure strategies. These choices are called "mixed strategies." A rigorous presentation of the mixed strategies is given in Appendix 1.

If the players use mixed strategies, then the capture time is a random variable, so that each player cannot guarantee a fixed cost but only an expected cost. Obviously, any pure strategy can be looked upon as a mixed strategy with degenerate probability distribution concentrated at that particular pure strategy, so that the pure strategies are included in the set of mixed strategies. A mixed strategy of the searcher will be denoted by s and a mixed strategy of the hider will be denoted by h. The expected cost of using the mixed strategies s and h will be denoted by $c(s, h)$. The maximal expected cost $v(s)$ of using a search strategy s,

$$v(s) = \sup_{h} c(s, h) = \sup_{H \in TH} c(s, H), \tag{5}$$

will be called "the value of strategy s," and the minimal expected cost $v(h)$ of using a hiding strategy h,

$$v(h) = \inf_{s} c(s, h) = \inf_{S \in TS} c(S, h), \tag{6}$$

will be called the "value of strategy h." It is obvious that for any s and h, $v(s) \geq v(h)$, because $v(s) \geq c(s, h) \geq v(h)$.

If there exists a real number v which satisfies

$$v = \inf_{s} v(s) = \sup_{h} v(h), \tag{7}$$

then we say that the game has a value v. In this case, for any $\varepsilon > 0$, there exist a search strategy $s_\varepsilon^*$ and a hiding strategy $h_\varepsilon^*$ which satisfy

$$v(s_\varepsilon^*) \leq (1 + \varepsilon)v \quad \text{and} \quad v(h_\varepsilon^*) \geq (1 - \varepsilon)v. \tag{8}$$

Such strategies will be called "$\varepsilon$-optimal strategies." In the case that there exists $s^*$ (resp. $h^*$) such that $v(s^*) = v$ (resp. $v(h^*) = v$), then $s^*$ (resp. $h^*$) is called an "optimal strategy."

In general, if the sets of pure strategies of both players are infinite, then the game need not have a value. (For details, see Luce and Raiffa (1957, Appendix 7).) However, in Appendix 1 we shall prove that any search game of the type already described has a value and an optimal search strategy. (The hider need not have an optimal strategy and in some games he has only $\varepsilon$-optimal strategies.)

Keeping the previous framework in mind, we present a general description of the search games considered in this monograph. In Part I we consider search games in compact spaces. In Chapter 1 we describe the framework of such games. In Chapter 2 we consider search games with an immobile hider in graphs and in multidimensional regions. In Chapters 3 and 4 we consider search games which take place in compact spaces, but here the problem is complicated by the fact that the hider can move in an arbitrary fashion.

In Part II we consider search games in some unbounded domains. The general framework of such problems is described in Chapter 5. In Chapter 6 we use the results presented in Appendix 2 in order to establish some general results on the optimality of the exponential functions for some minimax problems. The results obtained in Chapter 6 are used in Chapters 7 and 8, in which we consider search games on the infinite line, on a finite set of infinite rays, and some search games in the entire plane.

In Appendix 3 we describe some discrete search games which appear in the literature but do not fall into the framework of the main text.

Several search games are mentioned but not solved and some interesting open problems are presented. We hope that the material presented in this monograph will stimulate further research on some of these problems, and on search games in general.

# PART I
# SEARCH GAMES IN COMPACT SPACES

Chapter 1

# General Framework

The search spaces considered in Part I are closed and bounded subsets of a Euclidean space. They are usually either a compact region (i.e., the closure of a connected bounded open set) in a Euclidean space with two or more dimensions, or a graph. In this monograph, a "graph" will mean a connected set of arcs which can be of any type: a circle, a tree, a set of k arcs connecting two points, etc. Obviously, if a graph is given in the combinatorial form of vertices and edges, then it can be embedded in a three-dimensional Euclidean space $R^3$ in such a way that the edges intersect only at vertices of the graph. (Two dimensions are not sufficient for nonplanar graphs.) Thus, we shall look upon each graph as a subset of $R^3$. We shall use the following metric in the search space Q.

DEFINITION 1. The distance $d(Z_1, Z_2)$ between any two points $Z_1$, $Z_2$ is defined as the minimum length among all the paths which connect $Z_1$ and $Z_2$ and pass inside Q.

The diameter D of Q is defined as

$$D = \max_{Z_1, Z_2 \in Q} d(Z_1, Z_2).$$

A pure search strategy S is a continuous trajectory inside Q which does not exceed a fixed maximal velocity. The time unit will be chosen so as to normalize this maximal velocity to 1. Such a trajectory $S(t)$ is a mapping of the positive half of the real line, $0 \leq t < \infty$, into Q such that for any $0 \leq t_1 < t_2$,

$$d(S(t_2), S(t_1)) \leq t_2 - t_1.$$

We shall usually assume that the searcher has to start from a fixed point O to be called the "origin" (i.e., $S(0) = O$), but we shall sometimes consider other possibilities such as an arbitrary or a random starting point. The set of all pure search strategies is denoted by TS.

A pure hiding strategy H is an arbitrary continuous trajectory inside Q with maximal velocity not exceeding w. In the case $w = 0$, the hider is immobile and H is a single point, while if $w > 0$, then the hider is mobile and H is a trajectory which satisfies $d(H(t_1), H(t_2)) \leq w(t_2 - t_1)$ for all $t_2 > t_1 \geq 0$. The case of a mobile hider also includes the possibility of $w = \infty$, i.e., a hider with an unbounded velocity. The set of all pure hiding strategies is denoted by TH.

We assume that the searcher and the hider cannot see one another until their distance is less than or equal to the discovery radius r, and at that very moment capture occurs (and the game terminates). In cases where Q is a graph, then (for convenience) r will be taken as zero. (Actually, r can be chosen as a small positive number without introducing any significant changes in the results.) If Q is a multidimensional region, then it will be assumed that r is very small in comparison with the magnitude of Q.

In order to simplify the presentation of the results, we shall generally consider the case in which both the maximal velocity of the searcher and the radius of detection are constants. However, we shall also extend the results to the case where the maximal velocity of the searcher depends on his location and the radius of detection depends on the location of the hider. We shall call such a case "a nonhomogeneous search space."

We will often use the following notion of the maximal discovery rate which will enable us to present the results in a general form.

DEFINITION 2. The maximal discovery rate g of the searcher is defined a the maximal Lebesgue measure of a set which can be swept by the searcher in one unit of time.

Since the maximal velocity of the searcher is 1, it follows that g = 1 for the search in a graph. In case that Q is a two-dimensional region, the sweep width is 2r, so that the maximal area g of the strip which can be swept in one unit of time is 2r. By a similar reasoning, g is equal to $\pi r^2$ for three-dimensional regions, etc.

The capture time which is denoted by C(S, H) (and sometimes by T) represents the loss of the searcher (and the gain of the hider). It is formally defined as

$$\left.\begin{array}{l} C(S, H) = \min\{t : d(S(t), H(t)) \leq r\}. \\ \text{If no such } t \text{ exists, then } C(S, H) = \infty. \end{array}\right\} \qquad (1)$$

A mixed strategy s (resp. h) of the searcher (resp. hider) is a probability on TS (resp. TH). In order to rigorously present such strategies, one has to introduce a topology on TS and TH. Such a construction is presented in Appendix 1. We also show there that $C(S, H)$ is Borel measurable in both variables, so that we can define the payoff c, in the case that the searcher uses s and the hider uses h, as the expected value of C in the cross-product measure $s \times h$:

$$c(s, h) = \int C(S, H)\, d(s \times h). \tag{2}$$

The fundamental results proven in Appendix 1 are that any search game described above has a value v, i.e.,

$$\inf_s \sup_h c(s, h) = \sup_h \inf_s c(s, h), \tag{3}$$

and that the searcher always has an optimal strategy. Thus, for any such search game, the searcher can always guarantee an expected payoff not exceeding v, while the hider can guarantee that the expected payoff exceeds $(1 - \varepsilon)v$.

For the search games presented in this monograph, we shall generally use constructive methods to find the value and the optimal (or $\varepsilon$-optimal) strategies of the players. In the case of a graph, whenever we shall be able to obtain a solution of the game, it will be an exact solution. On the other hand, the solutions that we get for the search games in multi-dimensional regions depend upon the fact that the detection radius r is small. In this case, we shall present two strategies $s_\varepsilon^*$ and $h_\varepsilon^*$ and a function $f(r)$ which satisfy (see (5) and (6) in the Introduction) $v(s_\varepsilon^*) \leq (1 + \varepsilon)f(r)$ and $v(h_\varepsilon^*) \geq (1 - \varepsilon)f(r)$, where $\varepsilon \to 0$ as $r \to 0$. Thus, $s_\varepsilon^*$ and $h_\varepsilon^*$ are $\varepsilon$-optimal strategies and $v \sim f(r)$ for small r.

In calculating the expected capture time of the search games to be considered, we shall often use the following result, which is well known in probability theory (see, e.g., Feller (1971, p. 150)).

PROPOSITION. Let T be a nonnegative random variable with distribution function F ($F(t) = \Pr(T \le t)$). Then the expected value of T satisfies

$$ET = \int_0^\infty \Pr(T > t)\, dt. \tag{4}$$

Proof. Let $I(t, T)$ be the function

$$I(t, T) = \begin{cases} 1, & 0 \le t < T, \\ 0 & \text{otherwise.} \end{cases}$$

Then

$$ET = \int_0^\infty T\, dF(T) = \int_0^\infty \int_0^\infty I(t, T)\, dt\, dF(T)$$

$$= \int_0^\infty \int_0^\infty I(t, T)\, dF(T)\, dt \qquad \text{(by Fubini's theorem)}$$

$$= \int_0^\infty (1 - F(t))\, dt = \int_0^\infty \Pr(T > t)\, dt. \qquad \text{Q.E.D.}$$

Since $\Pr(T > t)$ is monotonic nonincreasing in t, it follows from (4) that for any positive number $\beta$

$$ET = \sum_{i=0}^\infty \int_{i\beta}^{(i+1)\beta} \Pr(T > t)\, dt \le \beta \sum_{i=0}^\infty \Pr(T > i\beta). \tag{5}$$

Similarly,

$$ET \ge \beta \sum_{i=0}^\infty \Pr(T > (i + 1)\beta) = \beta \sum_{i=1}^\infty \Pr(T > i\beta). \tag{6}$$

We now present a simple but useful result which will enable us to normalize the arc lengths in some networks, and will also be used for search games in unbounded domains. It actually states that changing the length unit in Q affects the search game in a very simple manner.

SCALING LEMMA. Let G be a search game in a set Q with an origin O and a detection radius r. Assume that the value of G is v and that $s^*$, $h^*$ are optimal ($\varepsilon$-optimal) strategies.

Consider a set $\overline{Q}$ which is obtained from Q by a mapping $\varphi$ with the following property. There exists a positive constant $\alpha$ such that for all $Z_1, Z_2 \in Q$

$$\overline{d}(\varphi(Z_1), \varphi(Z_2)) = \alpha\, d(Z_1, Z_2) \tag{7}$$

($\overline{d}$ is the distance in $\overline{Q}$).

Define a search game $\overline{G}$ in $\overline{Q}$ with an origin $\overline{O} = \varphi(O)$, a detection radius $\overline{r} = \alpha r$, and the same maximal velocities for the searcher and the hider as in G. Then the value $\overline{v}$ of $\overline{G}$ satisfies $\overline{v} = \alpha v$ and the optimal ($\varepsilon$-optimal) strategies of $\overline{G}$ are obtained by applying the mapping $\varphi$ to the trajectories in Q and changing the time scale by a factor of $\alpha$.

Proof. Let $\overline{TS}$ and $\overline{TH}$ be the sets of admissible trajectories of $\overline{G}$. Define the following mapping $\phi$ of TS into $\overline{TS}$ and of TH into $\overline{TH}$:

$$\overline{S} = \phi(S) \quad \text{iff} \quad \overline{S}(t) = \varphi(S(t/\alpha)), \qquad 0 \leq t < \infty,$$

and

$$\overline{H} = \phi(H) \quad \text{iff} \quad \overline{H}(t) = \varphi(H(t/\alpha)), \qquad 0 \leq t < \infty.$$

At first we show that $\overline{S}$ is an admissible trajectory. This follows from the fact that for any $t_1$, $t_2$,

$$\overline{d}(\overline{S}(t_1), \overline{S}(t_2)) = \alpha\, d(S(t_1/\alpha), S(t_2/\alpha)) \qquad \text{(by (7))}$$

$$\leq \alpha \left| \frac{t_1}{\alpha} - \frac{t_2}{\alpha} \right| = |t_1 - t_2|.$$

A similar argument holds for $\overline{H}$.

Similarly, it can be shown that $\phi^{-1}$ maps each admissible trajectory $\overline{S}$ (resp. $\overline{H}$) to an admissible S (resp. H). Thus the mapping $\phi$ is one to one and onto.

Next note that the payoff $\overline{C}$ of the game $\overline{G}$ satisfies for all S and H

$$\begin{aligned}\overline{C}(\phi(S), \phi(H)) &= \min\{t : \overline{d}(\overline{S}(t), \overline{H}(t)) \leq \alpha r\} \\ &= \min\{t : \overline{d}(\varphi(S(t/\alpha)), \varphi(H(t/\alpha))) \leq \alpha r\} \\ &= \min\{t : d(S(t/\alpha), H(t/\alpha)) \leq r\} = \alpha C(S, H).\end{aligned}$$

Now, the mapping $\phi$ can be extended to include mixed strategies by defining $\overline{s} = \phi(s)$ iff for any $A \subset TS$, the probability measure of A under s is equal to the probability measure of $\phi(A)$ under $\overline{s}$ and similarly defining $\overline{h} = \phi(h)$. It is easy to see that $\phi$ is one to one and onto and that $\overline{c}(\phi(s), \phi(h)) = \alpha c(s, h)$ for all s, h. This implies that if $s^*$, $h^*$ are optimal ($\varepsilon$-optimal) for G, then $\phi(s^*)$, $\phi(h^*)$ are optimal ($\varepsilon$-optimal) for $\overline{G}$, and that $\overline{v} = \alpha v$. Q.E.D.

Chapter 2

# Search For An Immobile Hider

## 2.1 GENERAL RESULTS

In this chapter, we consider search games in compact spaces with an immobile hider. In this case, a pure hiding strategy H is a point in the search space Q, and a mixed hiding strategy h is a probability measure in Q. A pure search strategy S is a continuous trajectory in Q with maximal velocity not exceeding 1. Since the hider is immobile, it can be assumed that the searcher will always use his maximal velocity because any (pure) search strategy $S_1(t)$ which does not use the maximal velocity is dominated by a search strategy $S_2(t)$ which uses the maximal velocity along the trajectory traced by $S_1(t)$. A mixed search strategy s is a probability measure on the set of the above described trajectories.

A hiding strategy which plays an important role in some of the games to be presented is the completely randomized strategy $h_R$ which corresponds to a uniform probability distribution in Q. More precisely:

DEFINITION 1. The completely randomized strategy $h_R$ is a random choice of the hiding point H such that for all measurable sets $B \subset Q$,

$$\Pr(H \in B) = \mu(B)/\mu,$$

where $\mu(B)$ is the Lebesgue measure of B and $\mu$ the Lebesgue measure of Q.

In the case of a multidimensional region, the Lebesgue measure of B, or of Q, has its usual meaning (area, volume, etc.). In the case that Q is a graph, $\mu$ simply means the sum of the lengths of the arcs which compose Q, and $\mu(B)$ is defined as the sum of the Lebesgue measures of the intersections of B with each of these arcs.

Using the notion of the maximal discovery rate g of the searcher (see Definition 2 of Chapter 1), we now derive a lower bound for the expected evasion time guaranteed by the hider if he uses the completely randomized strategy $h_R$.

LEMMA 1. For any search space, $v(h_R) \geq \mu/2g$.

Proof. It follows from Definition 2 of Chapter 1 that for any search trajectory S, for all $t > 0$, the Lebesgue measure of the set swept by S in the time segment $(0, t]$ does not exceed gt. Thus, if the hider uses $h_R$, then the probability that the discovery time $C(S, H)$ exceeds t is greater than or equal to $1 - (gt/\mu)$.

Since the discovery time is a nonnegative random variable, it follows from equality (4) of Chapter 1 that the expected discovery time, $c(S, h_R)$, satisfies

$$c(S, h_R) = \int_0^\infty \Pr(C(S, h_R) > t)\, dt$$

$$\geq \int_0^\infty \max(1 - (gt/\mu), 0)\, dt = \mu/2g. \qquad \text{Q.E.D.}$$

The following corollaries are immediate consequences of the considerations used in the proof of Lemma 1.

COROLLARY 1. If S satisfies $c(S, h_R) = \mu/2g$, then for all $0 \leq t \leq \mu/g$ the measure of the points swept by S in the time interval $(0, t]$ is equal to $gt$ (i.e., S sweeps without overlapping).

COROLLARY 2. Any search game which satisfies $v = \mu/2g$ would have the same value and the same optimal strategies even if we allow the searcher an arbitrary starting point, instead of the usual assumption $S(0) = O$.

The extension of the preceding discussion to search games with more than one searcher is presented in the following remark.

Remark 1. Consider a search game with one immobile hider and J searchers, with the j-th searcher having a maximal velocity $\nu_j$. Assume that all the searchers cooperate in order to discover the hider (by at least one of them) as soon as possible. Let $g_j$ be the Lebesgue measure of a set which can be swept by the j-th searcher in one unit of time, and define the total rate of discovery $\bar{g}$ by $\bar{g} = \sum_{j=1}^{J} g_j$. It should be noted that Lemma 1 remains valid for this case, with $\bar{g}$ replacing g.

Lemma 1 shows that $\mu/2g$ is a lower bound for the value of any search game. Note that in proving Lemma 1 we did not use the fact that the searcher moves along a continuous trajectory, and, in fact, this result is valid even in the case that the searcher can "jump" from one point to another in $Q$, as long as the basic requirement that the maximal discovery rate is $g$ holds.

Now suppose that we remove the requirement that the searching trajectory should be continuous and define a new game, still with an immobile hider, which we call the "unrestricted game." In this game, for each small time interval $\Delta t$, the searcher can choose a subset $B$ of $Q$ with $\mu(B) \leq g\, \Delta t$. If the hider is in $B$, then capture occurs at the middle of this time interval. It is easy to see that in the unrestricted game the searcher can make the expected discovery time equal to $\mu/2g$ by dividing the set $Q$ into $n$ disjoint subsets, each of them having measure $\mu/n$, and visiting them (without repetition) in a random order (i.e., choosing each permutation of these sets with probability $1/n!$). Since $\Delta t = \mu/ng$, it follows that the expected capture time for this strategy is

$$\sum_{i=1}^{n} \frac{(i - 1/2)\mu}{ng} \frac{1}{n} = \frac{\mu}{2g}.$$

From the preceding discussion, we conclude that the value of the unrestricted game is $\mu/2g$.

Remark 2. The unrestricted game is similar to a discrete search game in which $Q$ consists of $n$ cells of equal size. We now formulate and solve a more general discrete version of the search game. In the game to be considered, $Q$ consists of $n$ cells of sizes $\mu_1, \ldots, \mu_n$ and the measure of $Q$ is defined as $\mu = \sum_{i=1}^{n} \mu_i$. It is assumed that the maximal rate of discovery of the searcher is $g$, so that it takes him $\mu_i/g$ units of time to look at cell number $i$. It is also assumed that if the hider is located in cell i and if the searcher starts to look in this cell at time t, then the hider is discovered at time $t + \mu_i/2g$.

A pure hiding strategy H is an element of the set $\{1, 2, \ldots, n\}$, while a pure searching strategy S is a permutation $(i_1, \ldots, i_n)$ of the numbers $(1, 2, \ldots, n)$. We now show that the result $v = \mu/2g$ also holds for this discrete version.

PROPOSITION. The value of the discrete search game is $\mu/2g$. An optimal hiding strategy $h^*$ assigns a probability of $\mu_i/\mu$ to each cell, and an optimal search strategy $s^*$ is to take any permutation $(i_1, \ldots, i_n)$ and to assign a probability 1/2 to $(i_1, \ldots, i_n)$ and a probability 1/2 to its "reverse" $(i_n, \ldots, i_1)$.

<u>Proof</u>. The strategy $h^*$ satisfies for any permutation $S = (i_1, \ldots, i_n)$

$$c(S, h^*) = \sum_{j=1}^{n}\left(\left(\sum_{m=1}^{j-1} \mu_{i_m}\right) + \mu_{i_j}/2\right)\mu_{i_j}/\mu g$$

$$= \left(\sum_{1\le m\le j\le n} \mu_{i_m}\mu_{i_j} + \frac{1}{2}\sum_{j=1}^{n} \mu_{i_j}^2\right)/\mu g$$

$$= \left(\sum_{i=1}^{n} \mu_i\right)^2/2\mu g = \mu/2g,$$

while $s^*$ satisfies for all $H\epsilon\{1, \ldots, n\}$

$$c(s^*, H) = (C((i_1, \ldots, i_n), H) + C((i_n, \ldots, i_1), H))/2 = \mu/2g.$$

Thus, $v(h^*) = v(s^*) = v = \mu/2g$. Q.E.D.

A brief description of other discrete search games which appear in the literature is given in Appendix 3.

The expression $\mu/2g$ can be looked upon as the value that is obtained if the searcher is able to carry out his search with maximal efficiency. The games considered in this monograph are obviously restricted by the fact that the searcher has to move along a continuous trajectory so that the value does depend on the structure of Q. We shall have cases, such as Eulerian graphs, in which the searcher can perform the search with "maximal efficiency" which assures him a value of $\mu/2g$. A similar result holds for search in two-dimensional regions with a small detection radius, and in this case the searcher can keep the expected discovery time below $(1 + \varepsilon)\mu/2g$. On the other hand, in the case of a non-Eulerian graph, we shall prove that the value is greater than $\mu/2g$. We shall also show that the maximal value is $\mu/g$. This value is obtained in the case that Q is a tree. We shall also demonstrate the complications encountered in finding the optimal strategies in the case that Q is neither an Eulerian graph nor a tree.

Some of the results presented in Chapter 2 appear in a paper by Gal (1979).

## 2.2 SEARCH IN A GRAPH

In our discussion, the notion "graph" will mean any finite connected set of arcs in a three-dimensional Euclidean space which intersect only at vertices of Q. (Note that we allow more than one arc to meet the same pair of vertices.) In contrast to the common use in graph theory, the lengths of the arcs will be crucial. The sum of the lengths of the arcs in Q will be denoted by $\mu$.

In studying search trajectories in Q, we shall often use the notion "closed trajectory" defined as follows.

DEFINITION 2. A trajectory $S(t)$ defined for $0 \leq t \leq \tau$ is called "closed" if $S(0) = S(\tau)$. The line traced by this trajectory is called a "tour."

Note that a closed trajectory may cut itself and may even go through some of the arcs more than once.

We now consider a family of graphs which lend themselves to a simple solution of the search game. These are the Eulerian graphs defined as follows.

DEFINITION 3. A graph Q is called "Eulerian" if there exists a tour L with length $\mu$, which visits all the points of Q. The tour L will be called an "Eulerian tour."

It is well known that Q is Eulerian if and only if the degree (i.e., the number of arcs attached) of all the vertices of Q is even. (See, e.g., Berge, 1973.)

Since the maximal rate g of discovery in graphs is 1, it follows from Lemma 1 that $\mu/2$ is a lower bound for the value of the search game in any graph. We now show that this bound is attained if and only if Q is Eulerian.

THEOREM 1. The value of a search game in a graph is equal to $\mu/2$ if and only if Q is an Eulerian graph.

Proof. If Q is an Eulerian graph, then the searcher can guarantee an expected discovery time not exceeding $\mu/2$ by using the strategy $s^*$ described as follows. Pick an Eulerian tour L to be traced by the searcher and choose each direction of encircling L with probability 1/2.

Since the length of L is $\mu$, it follows that for all $H \in Q$, $c(s^*, H) = \mu/2$ so that $v(s^*) = \mu/2$. On the other hand, if the hider uses the completely randomized strategy $h_R$ (see Definition 1), then it follows from Lemma 1 that $v(h_R) \geq \mu/2$. Since for any s, h, $v(s) \geq v \geq v(h)$, it follows that $v = v(s^*)$ $= v(h_R) = \mu/2$, so that $s^*$ and $h_R$ are optimal strategies.

We now show that if $v = \mu/2$, then Q is Eulerian. If $v = \mu/2$, then there exists a search strategy $s^*$ such that, for all $H \in Q$, $c(s^*, H) \leq \mu/2$. On the other hand, it follows from Lemma 1 that all search strategies s satisfy

$$c(s, h_R) \geq \mu/2. \tag{1}$$

Thus

$$c(s^*, h_R) = \mu/2, \tag{2}$$

so that

$$c(s^*, H) = \mu/2 \quad \text{for almost every } H. \tag{3}$$

It follows from (1) and (2) that almost every trajectory S used by $s^*$ satisfies $c(S, h_R) = \mu/2$, so that Corollary 1 of Lemma 1 implies that for all $0 < t \leq \mu$, the set $\{S(\tau), 0 < \tau \leq t\}$ has measure t. Thus, S goes along a trajectory which passes through each arc in Q only once; such an S will be called an "Eulerian path."

Let $H_1$ and $H_2$ be two distinct points, on the same arc b, which satisfy (by (3))

$$c(s^*, H_1) = c(s^*, H_2) \qquad (= \mu/2). \tag{4}$$

Since $s^*$ uses Eulerian paths with probability 1, it follows that almost every trajectory used by $s^*$ contains a part which either goes from $H_1$ to $H_2$ along b or from $H_2$ to $H_1$ along b, and it follows from (4) that the probability of movement in each of these directions is 1/2. Thus for any arc b in Q there is probability of 1/2 that the search trajectory will move along b in each direction.

Since Q contains Eulerian paths, it follows that if Q is not Eulerian, then Q has a vertex A, with an odd degree, such that any Eulerian trajectory ends at A. Denote the arcs which are attached to A by $b_1, \ldots, b_n$, where $n = 2k + 1$ and k is an integer. Let $P_i$, $i = 1, \ldots, n$, be the probability that the searcher will move in $b_i$ toward A. It follows from the previous discussion that for all $i = 1, \ldots, n$, $P_i = 1/2$; on the other hand, almost all the trajectories used by $s^*$ are Eulerian paths ending at A, which means that they move along $k + 1$ of these arcs toward A and they move in the opposite direction for the other k arcs. Let $J_i = 1$ if the searcher moves along $b_i$ toward A and $J_i = 0$ otherwise. Then the previous discussion implies that for almost any S used by $s^*$, $\sum_{i=1}^{n} J_i = k + 1$, so that

$$\sum_{i=1}^{n} P_i = \sum_{i=1}^{n} EJ_i = E \sum_{i=1}^{n} J_i = k + 1$$

which contradicts the previous conclusion that all $P_i$ are equal to 1/2.

Thus, Q has to be Eulerian. Q.E.D.

Remark 3. It follows from Corollary 1 of Lemma 1 that any optimal search strategy in an Eulerian graph is a probability mixture of closed Eulerian trajectories.

Remark 4. For an Eulerian graph, the optimal strategies and the value of the search game remain the same even if instead of the usual assumption $S(0) = O$ we allow the searcher an arbitrary starting point. This result follows from Corollary 2 of Lemma 1.

The same result holds for the two-dimensional search game considered in Section 2.5. On the other hand, in the case of a non-Eulerian graph, the value of the game will usually decrease if we allow the searcher to choose his starting point.

We now establish an upper bound for the value, which holds for all graphs.

DEFINITION 4. A closed trajectory which visits all the points of Q and has minimal length will be called a "minimal tour" and denoted by $L^*$. Its length will be denoted by $\mu^*$.

LEMMA 2. Any minimal tour satisfies

$$\mu^* \leq 2\mu.$$

Proof. Consider a graph $\overline{Q}$ obtained from Q as follows. To any arc $b_i$ in Q, we add another arc $\overline{b}_i$ which connects the same vertices and has the same length as $b_i$. Obviously, any vertex in $\overline{Q}$ has an even degree, which implies that there exists an

Eulerian tour $\bar{L}$ in $\bar{Q}$ with length $2\mu$. If we now map the graph $\bar{Q}$ into Q such that both arcs $b_i$ and $\bar{b}_i$ of $\bar{Q}$ are mapped into one arc $b_i$ of Q, then the curve $\bar{L}$ is mapped into a tour L which passes through all the points of Q and has a length of $2\mu$. Q.E.D.

Remark 5. It can be seen that if Q is not a tree, then $\mu^* < 2\mu$ because in this case Q contains at least one closed loop which can be used as a part of $L^*$.

Remark 6. Finding a minimal tour for a given graph is called the Chinese postman problem. This problem can be reformulated as follows. Find in the given graph Q a set of arcs of minimum total length, such that when these arcs are duplicated (i.e., traversed twice), the degree of each vertex becomes even. This problem was solved by Edmonds (1965) and Edmonds and Johnson (1973) using a matching algorithm which uses $O(n^3)$ computational steps, n being the number of vertices in Q. This algorithm can be described as follows. First compute the shortest paths between all pairs of odd-degree vertices of Q. Then partition the odd-degree vertices into pairs so that the sum of lengths of the shortest paths joining the pairs is minimal. This can be done by solving a weighted matching problems. The arcs of Q in the paths identified with arcs of the matching are the arcs which should be duplicated (i.e., traversed twice). The algorithm is also described by Christofides (1975) and Lawler (1976).

Once an Eulerian graph is given, one can use the following simple algorithm for finding an Eulerian tour (see Berge, 1973). Begin at any vertex A and take any arc not yet used as long as removing this arc from the set of unused arcs does not disconnect the graph consisting of the unused arcs and incident vertices to them. Some algorithms which are more efficient than this simple algorithm were presented by Edmonds and Johnson (1973).

Actually, it is possible to slightly modify Edmond's algorithm in order to obtain a trajectory (not necessarily closed) which visits all the points of Q and has minimal length. This trajectory is a minimax search trajectory, and its length is the pure value of the game.

Using the length $\mu^*$ of the minimal tour, we now derive an upper bound for the value of the search game in a graph.

LEMMA 3. $v \leq \mu^*/2$.

Proof. If the searcher uses a strategy which encircles $L^*$ with probability 1/2 for each direction, then he guarantees an expected discovery time not exceeding $\mu^*/2$. (Note that a similar search strategy was used for Eulerian graphs, but in those cases it was an optimal strategy while for other graphs it is usually not an optimal strategy.) Q.E.D.

Combining Lemmas 1-3, we obtain:

THEOREM 2. For any graph, the value v of the search game with an immobile hider satisfies

$$\mu/2 \leq v \leq \mu^*/2 \leq \mu.$$

The lower bound is attained if and only if Q is an Eulerian graph. In the next section, we show that the upper bound $\mu$ is attained in the case that Q is a tree.

Remark 7. Theorem 2 brings to mind the rendezvous value of a metric space introduced by Gross (1964). For a compact connected metric space, he considered the two-person zero-sum game in which each person independently chooses a point in this space and the payoff is given by the distance between them. Such a game has a value $v$ which satisfies $D/2 \leq v \leq D$, where $D$ is the diameter of the space. Using this result, Gross showed that relative to such a space there exists a unique constant $\alpha$ ($\alpha = v$) with the property that given any finite collection of points, one can find a point $Z$ such that the average distance of the points in the collection from $Z$ is equal to $\alpha$.

## 2.3 SEARCH ON A TREE

In this section, we consider the case when the set $Q$ is a tree. By adding an extra node at the origin $O$ (if it is not a node in the original tree), the tree can always be represented so that the origin (i.e., the starting point) is at the root of the tree. We shall establish the following result.

THEOREM 3. Let $v$ be the value of the search game for an immobile hider in a tree $Q$. Then

$$v = \mu. \tag{5}$$

Proof. It follows from Theorem 2 that $v \leq \mu$. We shall establish the fact that $v \geq \mu$ by proving the following lemma.

LEMMA 4. Consider two trees $Q$ and $Q'$ as depicted in Figure 1. The only difference between $Q$ and $Q'$ is that two adjacent terminal branches $BA_1$ of length $a_1$ and $BA_2$ of length $a_2$

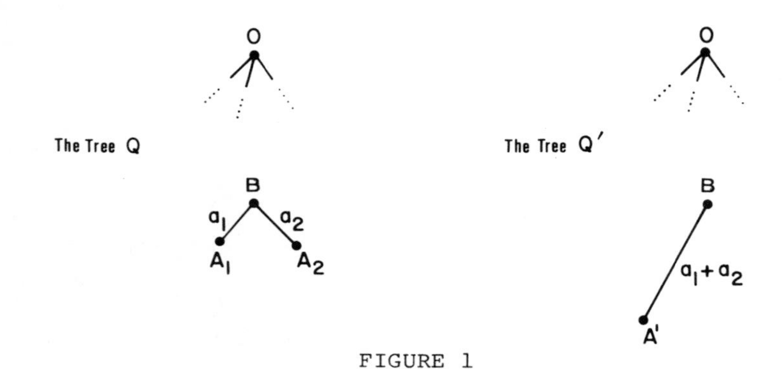


FIGURE 1

are replaced by one terminal branch BA′ of length $a_1 + a_2$. Let v be the value of the search game in Q and let v′ be the value of the search game in Q′. Then $v \geq v'$.

Proof of Lemma 4. Any hiding strategy in a tree is obviously dominated by a hiding strategy which chooses points only among the terminal nodes of the tree, and thus we can assume that an optimal hiding strategy $h'^*$ has this property. Such a hiding strategy $h'^*$ satisfies for any search trajectory S′ in Q′

$$c(S', h'^*) \geq v'. \tag{6}$$

Let us define a hiding strategy $h^*$ in Q by attaching to each terminal node of Q the same probability as in $h'^*$, except that the probability of choosing the new nodes $A_1$ and $A_2$, which will be denoted by $p_1$ and $p_2$, respectively, is given by

$$p_1 = \frac{a_1}{a_1 + a_2} p' \quad \text{and} \quad p_2 = \frac{a_2}{a_1 + a_2} p', \tag{7}$$

where p′ is the probability of choosing A′ under $h'^*$ and $a_1$, $a_2$ are depicted in Figure 1.

We shall show that $v \geq v'$, by proving that for any search trajectory S in Q

$$c(S, h^*) \geq v'. \tag{8}$$

In order to prove (8) we proceed as follows. If the hider uses a strategy $h^*$ which chooses its hiding point at terminal nodes only, then it is best for the searcher to use only a search trajectory which has the following characteristics. It starts from the root O, moves in the shortest route to a terminal node, then moves in the shortest route to another terminal node, and so on until all the terminal nodes have been visited. Therefore, there is a one-to-one correspondence between the relevant search trajectories and the permutations of the terminal nodes. Bearing that in mind, and assuming (for convenience) that the search strategy S visits the terminal node $A_1$ before visiting $A_2$ (the proof is similar for the case where $A_2$ is visited before $A_1$), S can be represented by the following permutation of terminal nodes:

$$S \sim (A_{i_1}, \ldots, A_{i_I}, A_1, A_{j_1}, \ldots, A_{j_J}, A_2, A_{l_1}, \ldots, A_{l_L}).$$

Let us denote the distance from the origin O to the terminal node $A_k$ along the trajectory S, by $d_k$ and let $h^*$ assign a probability $p_k$ to the hiding point $A_k$. Then inequality (8) is equivalent to

$$\sum_{m=1}^{I} d_{i_m} p_{i_m} + d_1 p_1 + \sum_{m=1}^{J} d_{j_m} p_{j_m} + d_2 p_2 + \sum_{m=1}^{L} d_{l_m} p_{l_m} \geq v'. \tag{9}$$

In order to prove (9), let us consider two search trajectories in $Q'$:

$$S_1' \sim (A_{i_1}, \ldots, A_{i_I}, A', A_{j_1}, \ldots, A_{j_J}, A_{l_1}, \ldots, A_{l_L}) \qquad (10)$$

and

$$S_2' \sim (A_{i_1}, \ldots, A_{i_I}, A_{j_1}, \ldots, A_{j_J}, A', A_{l_1}, \ldots, A_{l_L}). \qquad (11)$$

It follows from (6) that

$$c(S_1', h'^*) \geq v' \qquad (12)$$

and

$$c(S_2', h'^*) \geq v'. \qquad (13)$$

Let us denote the distances from the origin O to the terminal node $A_k$ along the trajectories $S_1'$ and $S_2'$ by $d_{1k}$ and $d_{2k}$, respectively, and the distances to the terminal node $A'$ by $d_1'$ and $d_2'$, respectively. The following relations hold:

$$\begin{aligned} d_{1i_m} &= d_{2i_m} = d_{i_m}, & d_{1j_m} &= d_{j_m} + 2a_2, \\ d_1' &= d_1 + a_2, & d_{2j_m} &\leq d_{j_m} - 2a_1, \\ d_2' &\leq d_2 - a_1, & d_{1l_m} &\leq d_{l_m}, \\ & & d_{2l_m} &\leq d_{l_m}. \end{aligned} \qquad (14)$$

It follows from (12) and (13) that

$$\frac{a_1}{a_1 + a_2} c(S_1', h'^*) + \frac{a_2}{a_1 + a_2} c(S_2', h'^*) \geq v'.$$

Thus

$$\frac{a_1}{a_1 + a_2}\left(\sum_{m=1}^{I} d_{1i_m} p_{i_m} + d_1' p' + \sum_{m=1}^{J} d_{1j_m} p_{j_m} + \sum_{m=1}^{L} d_{1l_m} p_{l_m}\right)$$

$$+ \frac{a_2}{a_1 + a_2}\left(\sum_{m=1}^{I} d_{2i_m} p_{i_m} + \sum_{m=1}^{J} d_{2j_m} p_{j_m} + d_2' p' + \sum_{m=1}^{L} d_{2l_m} p_{l_m}\right) \geq v'.$$

Using (14), we obtain

$$\sum_{m=1}^{I} d_{i_m} p_{i_m} + \sum_{m=1}^{J} d_{j_m} p_{j_m} + \sum_{m=1}^{L} d_{l_m} p_{l_m} + \frac{a_1}{a_1 + a_2} d_1 p' + \frac{a_2}{a_1 + a_2} d_2 p' \geq v', \tag{15}$$

and now inequality (9) immediately follows from (7) and (15). Q.E.D.

Using Lemma 4, we can readily establish by induction on the number of terminal nodes that the inequality $v \geq \mu$ holds for any tree, thus completing the proof of Theorem 3.

It is worth noting that Lemma 4 also presents, as a by-product, a method for finding an optimal hiding strategy (an optimal strategy for the searcher is the one described in the proof of Lemma 3).

## 2.4 THE CASE OF AN ODD NUMBER OF ARCS CONNECTING TWO POINTS

In the case that the graph Q is neither Eulerian nor a tree, it follows from Theorem 2 and Remark 5 that $\mu/2 < v < \mu$. In this section, we consider an example for such a situation. In this example, the graph Q consists of k nonintersecting arcs, $b_1, \ldots, b_k$, of equal length, connecting two points O and A. An immediate consequence of the scaling lemma presented in

Chapter 1 is that it is sufficient to consider the case in which all the arcs have unit length. This example will also be considered in Chapter 3, where we deal with a mobile hider.

If the number $k$ of arcs is even, then the graph $Q$ is Eulerian and the solution of the game is simple. On the other hand, it turns out that the solution is surprisingly complicated in the case that $k$ is an odd number greater than 1, even if $k$ is equal only to 3.

It should be noted that in the cases of the Eulerian graphs and of the trees considred so far, the value $v$ is equal to $\mu^*/2$ (see Definition 4), which implies that the search strategy used to establish Lemma 3 is an optimal strategy. On the other hand, in cases that $v < \mu^*/2$, obviously, such a search strategy cannot be optimal. The next lemma demonstrates the fact that such a situation occurs in the search on an odd number of arcs.

LEMMA 5. If $Q$ is a set of $k$ nonintersecting arcs of unit length which join $O$ and $A$, and $k$ is an odd number greater than 1, then $v < \mu^*/2$.

Proof. Obviously, $\mu^* = k + 1$. Thus we have to show that

$$v < (k + 1)/2. \tag{16}$$

Before proving (16), we present the following definition which will be frequently used in the sequel.

DEFINITION 5. Let $G$ be a set consisting of $k$ elements. If we choose each element of $G$ with probability $1/k$, we shall say that we have performed an equiprobable choice in $G$.

We now proceed in establishing (16) by presenting the following search strategy $\bar{s}$. Start from $O$, make an equiprobable choice among the k arcs, and move along the chosen arc to A; then make an equiprobable choice among the k - 1 remaining arcs, independently of the previous choice, and move along this arc back to O; then move back to A; and so on until all the arcs have been visited.

Let H be any pure strategy (i.e., a point in Q) and assume that its distance from A is d so that its distance from O is 1 - d. Let $m = (k - 1)/2$, let $E_i$, $i = 1, \ldots, m$, be the event that the hider is discovered during the time period $(2(i - 1), 2i]$, and let $E_f$ be the event that the hider is discovered during the time period $(k - 1, k]$. Then

$$c(\bar{s}, H) = \sum_{i=1}^{m} (2i - 1) \Pr(E_i) + (k - d) \Pr(E_f)$$

$$= \sum_{i=1}^{m} (2i - 1) \cdot 2/k + (k - d)/k$$

$$= \frac{k}{2} + \frac{1}{2k} - \frac{d}{k} < \frac{k}{2} + \frac{1}{2k} < \frac{k + 1}{2}. \qquad \text{Q.E.D.} \qquad (17)$$

The case where Q consists of an odd number of arcs which connect two points has been used as an example for situations in which $v < \mu^*/2$, but this case is interesting by itself. It is amazing that the solution of the game is simple for any even k (and also, as will be demonstrated in the next chapter, for any odd or even k if the hider is mobile), but it may be quite complicated to solve this game even for the case that k is equal only to 3. The reasonable symmetric search strategy $\bar{s}$, used in (17), which is optimal for an even k, can assure the searcher of paying no more than $(k/2) + (1/2k)$, but we shall immediately show that $\bar{s}$ is not an optimal strategy for an odd number of arcs. (Incidentally, the fact that the search strategy $\bar{s}$ is not optimal, or even $\varepsilon$-optimal, can be easily deduced from the following argument. If the searcher uses $\bar{s}$, then the hider can guarantee a payoff which is close to $(k/2) + (1/2k)$ only by hiding near A with probability $1 - \varepsilon$. However, it can easily be verified that the payoff guaranteed by such a hiding strategy does not exceed $1 + \delta$, where $\delta$ is small. Thus, the value of the game has to be smaller than $(k/2) + (1/2k)$, which implies that the strategy $\bar{s}$ cannot be optimal, or even $\varepsilon$-optimal.)

In order to demonstrate the complexity of this problem, we now consider the case of $k = 3$. In this case, the symmetric search strategy $\bar{s}$ satisfies $v(\bar{s}) = 3/2 + 1/6 = 5/3$, but we now present a strategy[†] $s^*$ which satisfies

$$v(s^*) = (4 + \ln 2)/3 < 5/3.$$

The strategy $s^*$ is a specific choice among the following family $\{s_F\}$ of search strategies.

DEFINITION OF $\{s_F\}$. Consider a set of trajectories $S_{ij\alpha}$, where i and j are two distinct integers in the set {1, 2, 3} and $0 \leq \alpha \leq 1$. The trajectory $S_{ij\alpha}$ starts from O, moves along $b_i$ to A, moves along $b_j$ to the point $A_\alpha$ which has a distance of $\alpha$ from A (see Figure 2), moves back to A, moves to O along

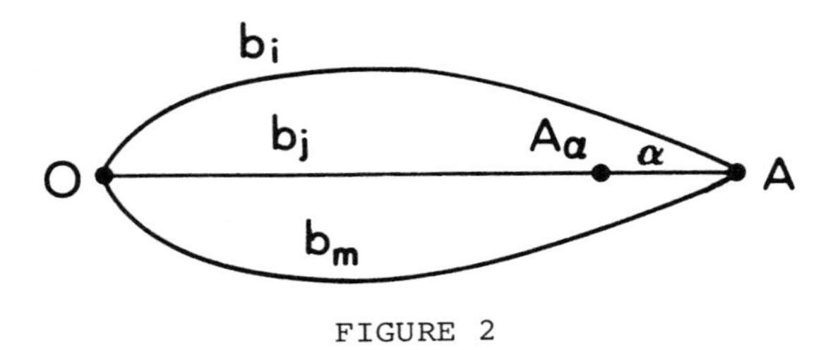


FIGURE 2

[†]This strategy was suggested by Donald J. Newman.

$b_m$, where $m \in \{1, 2, 3\} - \{i, j\}$, and then moves from O to $A_\alpha$ along $b_j$.

Let $F(\alpha)$ be a cumulative probability distribution function of a random variable $\alpha$ $(0 \le \alpha \le 1)$. Then the strategy $s_F$ is a probabilistic choice of a trajectory $S_{ij\alpha}$, where i is determined by an equiprobable choice in the set $\{1, 2, 3\}$, j is determined by an equiprobable choice in the set $\{1, 2, 3\} - \{i\}$, and $\alpha$ is chosen independently, using the probability distribution F.

Note that the symmetric strategy $\bar{s}$ is a member of the family $\{s_F\}$ with the random variable $\alpha$ being identically zero.

We now show that there exists a search strategy $s^* = s_{F^*}$ in $\{s_F\}$ with value less than $v(\bar{s})$. The distribution function $F^*$ of $\alpha$ which corresponds to this strategy is defined as

$$F^*(\alpha) = \begin{cases} 0, & \alpha < 0, \\ \frac{1}{2} + \frac{1}{4}e^\alpha, & 0 \le \alpha \le \ln 2, \\ 1, & \ln 2 < \alpha. \end{cases} \tag{18}$$

(Note that $F^*$ has a probability mass of 3/4 at $\alpha = 0$.) Let H be any hiding point and denote its distance from A by $\beta$. Then

$$c(s^*, H) = \frac{1}{3}(1 - \beta) + \frac{1}{3}\left(1 + 2\int_0^{\ln 2} \alpha \, dF^*(\alpha) + \beta\right)$$
$$+ \frac{1}{3}\left(1 + \int_0^{\beta} (2\alpha + 2 - \beta) \, dF^*(\alpha) + \beta(1 - F^*(\beta))\right).$$

It is easy to check that the derivative of $c(s^*, H)$ with respect to $\beta$ is equal to zero for $0 < \beta < \ln 2$ and to $-1/3$ for $1 > \beta > \ln 2$. Thus, it is sufficient to calculate $c(s^*, H)$ for $\beta = \varepsilon$, where $\varepsilon$ is small, and the calculation readily shows that

$$v(s^*) = (4 + \ln 2)/3 < 5/3 = v(\bar{s}), \tag{19}$$

where $\bar{s}$ is the symmetric strategy.

We now show that $s^*$ is the best search strategy in the family $\{s_F\}$ by presenting a hiding strategy $h^*$ which satisfies for all $s_F \in \{s_F\}$

$$c(s_F, h^*) \geq (4 + \ln 2)/3. \tag{20}$$

The hiding strategy $h^*$ is defined as follows. Make an equiprobable choice of an arc $b_i$, $i \in \{1, 2, 3\}$, and hide at the point of $b_i$ which has a distance $\beta$ from A, where $\beta$ is a random variable which has the probability density $2e^{-\beta}$ for $0 < \beta < \ln 2$ and zero otherwise.

It is easy to see that for all $S_{ij\alpha}$ used by any search strategy $s_F$

$$\frac{dc(S_{ij\alpha}, h^*)}{d\alpha} = 2\left(\frac{1}{3} + \frac{1}{3}\int_\alpha^{\ln 2} 2e^{-\beta}\, d\beta\right) - \frac{2}{3}2e^{-\alpha}$$

$$= 0 \quad \text{if} \quad 0 < \alpha < \ln 2$$

and

$$\frac{dc(S_{ij\alpha}, h^*)}{d\alpha} = \frac{2}{3} \quad \text{if} \quad \ln 2 < \alpha < 1.$$

Thus

$$c(s_F, h^*) \geq c(S_{ij0}, h^*)$$

$$= \frac{1}{3}\left(1 - \int_0^{\ln 2} 2\beta e^{-\beta}\, d\beta\right) + \frac{4}{3} = \frac{4 + \ln 2}{3}.$$

It follows that if the searcher can use only search trajectories $S_{ij\alpha}$, then $s^*$ and $h^*$ presented previously are optimal strategies and the value of the game is $(4 + \ln 2)/3$. It is very plausible that $s^*$ and $h^*$ are optimal even if the searcher is not restricted to $S_{ij\alpha}$ trajectories. In order to show that, one would have to show $c(S, h^*) \geq (4 + \ln 2)/3$ for all S (we have just proved this inequality for $S = S_{ij\alpha}$), but this task seems to involve many technical details.

## 2.5 SEARCH IN A MULTIDIMENSIONAL REGION

When considering search games with a mobile or an immobile hider in multidimensional compact regions, we would like to avoid unnecessary complications, and thus we shall make a rather weak assumption about the region Q. We shall use the following definition.

DEFINITION 6. A compact region Q is called "simple" if the boundary of Q can be represented as the union of two continuous single-valued functions in some coordinate system (Figure 3).

We shall usually make the following assumption.

ASSUMPTION 1. The compact region Q is the union of a finite number of simple regions with disjoint interiors.

Note that we allow multiply connected regions with a finite number of "holes" in them.

When considering search games in multidimensional regions, we assume that the hider is captured at the first instant in

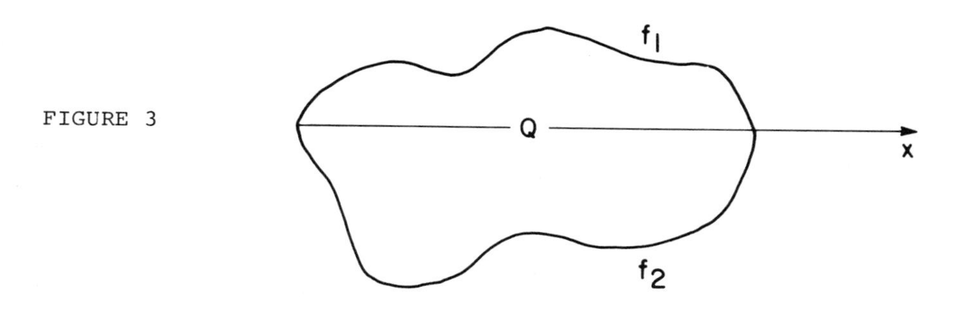


FIGURE 3

which the distance between him and the searcher is less than or equal to $r$, where the discovery radius $r$ is a small positive constant. (The case where $r$ is not a constant is considered in Section 2.6.)

In this section, we are concerned with an immobile hider. Since Lemma 1 holds for any search space $Q$, it follows that by using a completely randomized hiding strategy, the hider can guarantee that the expected discovery time is greater than or equal to $\mu/2g$, where $\mu$ is the Lebesgue measure of $Q$ and $g$ is the maximal discovery rate. Thus, the value $v$ of the game satisfies $v \geq \mu/2g$.

An upper bound for $v$ can be derived by introducing the notion of a "covering curve" defined as follows.

DEFINITION 7. A closed curve $L$ which passes inside $Q$ is called a "covering curve" if for any $Z \in Q$ there exists $Z' \in L$ such that $d(Z, Z') \leq r$.

It has been shown by Isaacs (1965, Section 12.3) that if $L$ is a covering curve with length $\tau$, then the searcher can guarantee an expected capture time not exceeding $\tau/2$ by choosing each one of the directions, of encircling $L$, with probability 1/2. (We used such a search strategy for Eulerian graphs.)

Now, if we could find a covering curve whose length $\tau$ satisfies

$$\tau \leq (1 + \varepsilon)\mu/g, \tag{21}$$

then we would have $v \sim \mu/2g$, and in this case the strategies of the searcher and of the hider already described would be $\varepsilon$-optimal. We now show that for any two-dimensional region which satisfies a rather weak condition, it is possible to find a covering curve whose length satisfies inequality (21).

LEMMA 6. Let Q be a two-dimensional compact region which satisfies Assumption 1 with the boundary of each one of the simple regions composing Q having a finte length. Then for any $\varepsilon > 0$ there is an $r_\varepsilon$ such that for any $r < r_\varepsilon$ there exists a covering vurve with length less than $(1 + \varepsilon)\mu/2r$.

<u>Proof</u>. Let $Q = \bigcup_{i=1}^{m} Q_i$, where $Q_i$ $1 \leq i \leq m$, is a simple region with area $\mu_i$. If for each $1 \leq i \leq m$ we could find a line which covers $Q_i$ with length less than $(1 + \varepsilon/2)\mu_i/2r$, then we could "link" these curves together and form a covering curve of Q by adding m arcs. The length of such a covering curve for Q would not exceed

$$(1 + \varepsilon/2)\mu/2r + mD = (1 + \varepsilon/2 + 2rmD/\mu)\mu/2r,$$

where D in the diameter of Q (see Definition 1 of Chapter 1). Obviously if r is small enough, then $2rmD/\mu < \varepsilon/2$, so that we would have the required covering curve for Q. Thus, we may assume that Q is already a simple region. Let $f_1$ and $f_2$ be the graphs of the continuous functions which bound Q. We can now cover Q by parallel strips of width 2r each, as depicted in Figure 4. Let $\gamma_1, \gamma_2, \ldots$ be the lengths of the vertical line segments formed by these strips. Then by moving alternately along these line segments and along the boundary of Q, we form a covering line for Q with length less than $\sum \gamma_j + \gamma$,

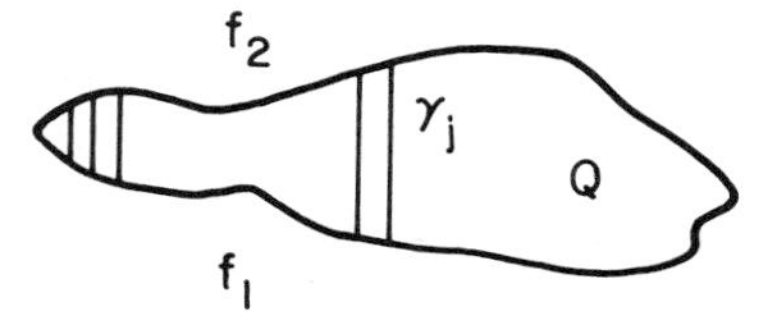


FIGURE 4

where $\gamma$ is the length of the boundary of Q. Due to the fact that $f_1$ and $f_2$ are Riemann integrable, it follows that for any $\varepsilon > 0$, if r is small enough, then $2r \sum \gamma_i < (1 + \varepsilon/2)\mu$. Thus, the length of the covering line is less than

$$(1 + \varepsilon/2 + 2r\gamma/\mu)\mu/2r < (1 + \varepsilon)\mu/2r$$

for a small enough r. Q.E.D.

The proof of Lemma 6 is based on the fact that two-dimensional regions can be covered by narrow strips with little overlap. The analog construction for three dimensions would require covering the region with narrow cylinders, but in this case the overlap would not be negligible. Thus, by using the completely randomized strategy, the hider can keep the expected capture time above $\beta\mu/g$, where $\beta > 1/2$. It should be noted, though, that in Chapter 4 we show that the value of a search game with a mobile hider in a multidimensional region satisfies $v < (1 + \varepsilon)\mu/g$, where $\varepsilon \to 0$ as $r \to 0$, and this bound is obviously applicable for the search game with an immobile hider.

Remark 8. The result of this section extends to the case of J searchers. Thus, if $\bar{g}$ is the total rate of discovery of the searchers (see Remark 1), then for two-dimensional regions $v \sim \mu/2\bar{g}$.

## 2.6 NONHOMOGENEOUS SEARCH SPACES

In this section, we assume that the search space is a multidimensional compact region of the same type considered in Section 2.5. However, here we allow the maximal velocity of the searcher, which we denote by $\nu(Z)$, to depend on the location of the searcher. We assume that $\nu(Z)$ is a continuous function which satisfies

$$0 < \nu_1 \leq \nu(Z) \leq \nu_2, \tag{22}$$

where $\nu_1$ and $\nu_2$ are constants. We also allow the detection radius to depend on the location of the hider. In this case, the detection radius is a function $r(Z)$ whereby if the hider is located at a point $Z$, then he can be seen from any point $Z_1$ which satisfies $d(Z_1, Z) \leq r(Z)$. We assume that

$$r(Z) = r_1\rho(Z), \quad \text{where } r_1 \text{ is a small number and } \rho(Z) \text{ is a continuous function which satisfies } \min_{Z\in Q} \rho(Z) > 0. \tag{23}$$

The discovery rate of the searcher, $g(Z)$, is defined as $2\nu(Z)r(Z)$ for two-dimensional regions, $\pi\nu(Z)r^2(Z)$ for three-dimensional regions, etc. The capture time $C(S, H)$ is given by

$$C(S,H) = \min\{t : d(S(t), H(t)) \leq r(H(t))\}.$$

Such a search space will be referred to as "a nonhomogeneous search space." The extension of Lemma 1 and the results established in Section 2.5 can be achieved using the following lemma:

LEMMA 7. Assume that $Q$ is a compact search space and that $\nu(Z)$ and $r(Z)$ satisfy (22) and (23). For any $\varepsilon > 0$, there exists a number $r_\varepsilon$ such that if $r_1 \leq r_\varepsilon$, then for all $Z$, $Z_1$ which satisfy $d(Z_1, Z) \leq r(Z)$,

$$(1 - \varepsilon)r(Z) < r(Z_1) < (1 + \varepsilon)r(Z) \tag{24}$$

and

$$(1 - \varepsilon)g(Z) < g(Z_1) < (1 + \varepsilon)g(Z). \tag{25}$$

Proof. It follows from (23) that $r(Z) = r_1\rho(Z)$ and $g(Z) = \text{constant} \cdot \tilde{\rho}(Z)$, where both $\rho(Z)$ and $\tilde{\rho}(Z)$ are positive continuous functions which does not depend on $r_1$. Thus, for any $\varepsilon > 0$, there exists a $\delta > 0$ such that for all $Z$, $Z_1$ which satisfy $d(Z_1, Z) \leq \delta$,

$$(1 - \varepsilon) < \frac{\rho(Z_1)}{\rho(Z)} < 1 + \varepsilon, \qquad (1 - \varepsilon) < \frac{\tilde{\rho}(Z_1)}{\tilde{\rho}(Z)} < 1 + \varepsilon. \tag{26}$$

Obviously, if $r_\varepsilon = \delta/\max_{Z\in Q} \rho(Z)$ and if $r_1 \leq r_\varepsilon$, then, by (23), $r(Z) \leq \delta$ for all $Z \in Q$. Thus, (26) implies that if $d(Z_1, Z) \leq r(Z)$, then (24) and (25) hold. Q.E.D.

We now extend Lemma 1 to the nonhomogeneous case. In this case, the natural randomization of the hider is given by choosing his location using probability density which is proportional to $1/g(Z)$.

LEMMA 1′. Let $h_1$ be the hiding strategy which uses the probability density $1/\tau g(Z)$, where

$$\tau = \int_Q \frac{d\mu(Z)}{g(Z)} \tag{27}$$

and $\mu$ is the Lebesgue measure. Then $h_1$ satisfies

$$v(h_1) \geq \frac{1}{1 + \varepsilon}\,\frac{\tau}{2}. \tag{28}$$

Proof. It follows from Lemma 7 that for any search trajectory $S$, the Lebesgue measure of the strip which is swept during a small time interval $\Delta t$ is less than $(1 + \varepsilon/2)g(S(t))\Delta t$. If the hider uses $h_1$, then the probability mass of such a strip is less than $(1 + \varepsilon)\ \Delta t/\tau$, where $\tau$ is given by (27). Thus, $\Pr(T \leq t) \leq (1 + \varepsilon)t/\tau$, so that

$$c(S, h_1) = \int_0^\infty \Pr(T > t)\ dt$$

$$\geq \int_0^{\infty} \max(0, 1 - (1 + \varepsilon)t/\tau)\, dt$$

$$= (1/(1 + \varepsilon))(\tau/2). \qquad \text{Q.E.D.}$$

We now extend the result established in Section 2.5 and show that the value of the search game with an immobile hider in a two-dimensional compact region satisfies $v \sim \tau/2$, where $\tau$ is given by (27) with $g(Z) = 2\nu(Z)r(Z)$. The fact that $v \geq (1/(1 + \varepsilon))(\tau/2)$, is an immediate consequence of Lemma 1′. An upper bound for the value can be obtained by constructing a covering curve (see Definition 7), such that the time required to encircle it is less than

$$(1 + \varepsilon)\int_Q \frac{d\mu(Z)}{2\nu(Z)r(Z)}.$$

This can be done by dividing Q into several regions $Q_1, \ldots, Q_n$ such that the variation of $\rho(Z)$ in $Q_i$, $i = 1, \ldots, n$, is small, and then using the technique adopted in Lemma 6.

Remark 9. All the results presented in Sections 2.2 and 2.3 can easily be extended to the case in which the maximal velocity of the searcher, $\nu(Z)$, depends on his location in the graph Q. (The radius of detection is assumed to be zero, as before.) In this case, the Lebesgue measure of Q, $\mu$, which appears in Theorems 1-3 should be replaced by

$$\tau = \int_Q \frac{d\mu(Z)}{\nu(Z)}.$$

Chapter 3

# Search For A Mobile Hider

## 3.1 GENERAL FRAMEWORK

The interest in search games with a mobile hider was aroused by the presentation of the princess and monster game described by Isaacs (1965, Section 12.4). In this game, the monster searches for the princess in a totally dark room Q (both of them being cognizant of its boundary). Capture occurs when the distance between the monster and the princess is less than or equal to r, where r is small in comparison with the dimension of Q. As a stepping stone for the general problem, Isaacs suggested a simpler problem in which Q is the boundary of a circle.

The princess and monster game on the boundary of a circle was solved several years later by Alpern (1974), Foreman (1974), and Zelikin (1972), under the additional assumption that the maximal velocity w of the hider, satisfies $w \geq 1$ (recall that the maximal velocity of the searcher is usually taken as 1). Their formulation is a little different from the one that we usually adopt, in the sense that the searcher does not start from a fixed point O. Instead, they assume that at $t = 0$ the initial relative position of the searcher with respect to the hider has a known probability distribution. (Alpern (1974) and Zelikin (1972) assume a uniform distribution, while Foreman (1974) considers a general probability distribution.) A discretized version of this problem was solved by Wilson (1972). An attempt to develop a technique for obtaining approximate solutions of some discrete games with a mobile hider by restricting the memory of the players was presented by Worsham (1974). Some versions of this game with a fixed termination time were considered by Foreman (1977).

A general solution of the princess and monster game in a convex multidimensional region, was presented by Gal (1979).[†] He also solved a generalization of the search on a circle by considering a search on k arcs which connect two points. Some of the results of Chapters 3 and 4 are taken from Gal (1979).

We now present the general framework of the search games with a mobile hider, to be considered in Chapters 3 and 4. We assume that the search space Q is either a graph or a compact multidimensional region. A pure strategy of the searcher is a continuous trajectory $S(t)$, $t \geq 0$, inside Q which satisfies

$S(0) = O$,

and for any $0 \leq t_1 < t_2$, $\quad d(S(t_1), S(t_2)) \leq t_2 - t_1$.

A pure strategy of the hider is a trajectory $H(t)$ $t \geq 0$ inside Q, with $H(0)$ an arbitrary point, which satisfies

for any $0 \leq t_1 < t_2$, $\quad d(H(t_1), H(t_2)) \leq w(t_2 - t_1)$,

where w is the maximal velocity of the hider.

[†]The material is taken with permission from SIAM J. control and Opt. 17 (1979), pp. 106-122. 

The capture time $T$ is the first instant $t$ with $d(S(t), H(t)) \leq r$, or infinity if no such $t$ exists. As usual, we take $r = 0$ for the search in a graph and assume that $r$ is a small positive number in the case that $Q$ is a multidimensional region.

When considering the role of mixed search strategies, one immediately observes that their benefit with respect to pure search strategies is much greater in the case of a mobile hider than in the case of an immobile hider. In order to see this fact, consider the case of an immobile hider in an Eulerian graph. In this case, by using a trajectory which traces an Eulerian curve, the searcher can guarantee an expected capture time not exceeding $\mu$, i.e., the pure value $\overline{VP}$ is equal to $\mu$, whereas the use of mixed strategies enables the searcher to guarantee an expected capture time not greater than $\mu/2$. Thus $v = \overline{VP}/2$, so that the use of mixed strategies against an immobile hider yields an improvement by a factor of 2. On the other hand, if a mobile hider is moving on a circle with maximal velocity $w \geq 1$, then no pure search strategy can guarantee capture, so that $\overline{VP} = \infty$. As we shall see in Section 3.3, the use of mixed search strategy enables the searcher to make sure that the expected capture time does not exceed $\mu$. Thus, the advantage of using mixed strategies is much greater in the case of a mobile hider.

Except for trivial cases, if only pure search strategies are used, then several searchers are needed to guarantee capture. The interesting problem of determining the minimal number of such searchers in a given graph, was considered by Parsons (1978a,b) and Megiddo and Hakimi (1978).

As in the discussion in Section 2.1, it is worthwhile to consider the unrestricted game (which is actually a discrete version of the search game) in which the physical constraint of the continuity of the trajectories of the players is disregarded. In this game, the set Q is divided into n "cells" $Q_1, \ldots, Q_n$, each of them having a measure $\mu/n$. The searcher and the hider can move from one cell into another at the time instants $t_i = i\ \Delta t$, $i = 0, 1, 2, \ldots$, where $\Delta t = \mu/ng$. The time interval $\Delta t$ may be regarded as the time required by the searcher to sweep one cell. We assume that capture occurs at the end of the first time interval in which both the searcher and the hider occupy the same cell.

It is easy to see that the value of the unrestricted game is $\mu/g$, because both the searcher and the hider can guarantee this value by using completely random strategies, choosing each cell with equal probability (1/n) and independently of previous choices, at each time instant $i\ \Delta t$, $i = 0, 1, 2, \ldots$ . It is worthwhile to estimate the probability of capture after time t in the unrestricted game, both in the case of a mobile compared to that of an immobile hider. If the hider is immobile, then, under optimal strategies, the probability of capture after time t satisfies $\Pr(T > t) \sim 1 - gt/\mu$ for $0 \leq t \leq \mu/g$, while if the hider is mobile, then $\Pr(T > t) \sim e^{-gt/\mu}$, $0 \leq t < \infty$.

In contrast to the unrestricted game, we shall be mainly concerned with games in which both the searcher and the hider are restricted to move along continuous trajectories, but it is interesting to note that in spite of that restriction, the value is still $(1 \pm \varepsilon)\mu/g$ for the games to be considered in Chapters 3 and 4. The search space of the game solved in

Section 3.2 consists of k arcs which connect two points. This game, which is an extension of the princess and monster game on the circle discussed in Section 3.3, serves as an introduction for the more difficult search game in a multi-dimensional region, which is solved in Chapter 4.

A remarkable property of the search games solved in Chapters 3 and 4 is the following. There exists a function $P(t)$, which decreases exponentially in $t$, such that for all $t$ both the searcher and the hider can keep the probability of capture after time $t$ around $P(t)$. In practice, this property may sometimes be more useful than the expected capture time, since the cost function need not be $T$ itself but may place a different (e.g., heavier) penalty on larger values of $T$. Indeed, we use this property to show that the optimal ($\varepsilon$-optimal) strategies obtained for the games in which the capture time serves as a cost function are still optimal ($\varepsilon$-optimal) even if we replace the capture time by a more general cost function. In other words, the optimal ($\varepsilon$-optimal) strategies of these games are uniformly optimal for all reasonable cost functions.

## 3.2 SEARCH ON k ARCS

In this section, we consider the following search game. The search space $Q$ is a set of $k$ nonintersecting arcs $b_1, \ldots, b_k$ joining two points $O$ and $A$ as depicted in Figure 1, the length of each arc being equal to 1. (The solution of this

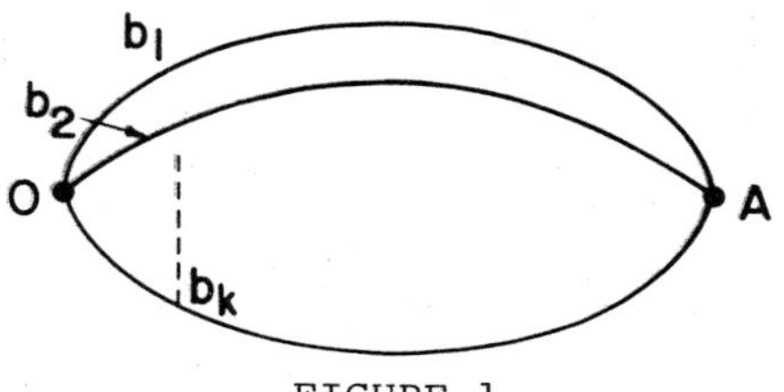


FIGURE 1

problem in the case that the length of the arcs is different from 1, is easily obtained by using the scaling lemma of Chapter 1.) The searcher has to start moving from O with maximal velocity equal to 1. The hider can choose an arbitrary starting point, and from this point he can move along any continuous trajectory in Q with maximal velocity w. In this chapter, we shall usually assume that $w \geq 1$. The capture time T, which is the loss of the searcher (or the gain of the hider), is the time elapsed until the searcher reaches a point which is occupied at the same time by the hider. In order to avoid unnecessary complications, we make the following assumption.

ASSUMPTION 1. The searcher can pass but not stay at the points O or A (or alternatively, the hider can pass from any arc $b_i$ to any other arc $b_j$ not only through O or A but also in an $\varepsilon$ distance from them). This assumption is not needed for $k = 2$ (search on a circle) as we shall see in Section 3.3.

We shall show that the optimal strategies of the searcher and of the hider are both of the type $R(Z, \tau)$ to be defined.

DEFINITION 1. Let Z be either the point O or the point A and let $0 \leq \tau \leq \infty$ be any real number. Then the (random) trajectory $R(Z, \tau)$ is defined by the following rules. At time $\tau$, starting from the point Z, choose an integer i equiprobably from the integers $\{1, 2, \ldots, k\}$ and move along the arc $b_i$ with unit velocity to the end point of this arc (O or A). Then (at time $\tau + 1$) make another equiprobable choice of an integer $j \in \{1, 2, \ldots, k\}$, independently of i, and move along $b_j$ until the other end point (A or O) is reached, and so on.

Using Definition 1, we state the following theorem.

THEOREM 1. For the search game on k arcs, an optimal strategy for the searcher is R(O, 0). An $\varepsilon$-optimal strategy for the hider is to start moving with maximal speed to the point A, stay there until time $1 - \varepsilon$, and then use the strategy $R(A, 1 - \varepsilon)$. The value of the game is k (which is equal to $\mu/g$).

The theorem will be proved using three lemmas. The following lemma is the fundamental one and will also be used in the next chapter.

LEMMA 1. Let Q be the set of k parallel arcs, each of length 1, joining O and A. Assume that at time $t = 0$, k horses leave the point O and each of them rides in unit velocity along a different arc from O to A. At the same time, $t = 0$, a giraffe starts moving from any point different from O along a trajectory H(t), where both the trajectory and the (not necessarily constant) velocity along this trajectory are arbitrary. The trajectory H(t) should be continuous so that in passing from any arc $b_1$ to any arc $b_j$ the giraffe has to go through either O or A. Then:

(a) For any trajectory H(t), the giraffe will meet at least one horse in the time period $0 < t \leq 1$.

(b) If the giraffe moves with velocity which does not exceed one, then he will not meet more than one horse in the time period $0 < t < 1$.

<u>Proof of (a)</u> Let $b_i(t)$, $i = 1,\ldots, k$, $0 \leq t \leq 1$, be the point located by the i-th horse at time t and let

$$G(t) = \{b_i(t),\ i = 1, 2,\ldots, k\}. \tag{1}$$

Then the set $G(t)$ is the boundary of the set

$$G^+(t) = \{b_i(u),\ t < u \leq 1,\ i = 1, 2, \ldots, k\}. \tag{2}$$

$G^+(0)$ is the whole set $Q$ except for the point $O$ while $G^+(1)$ is empty.

Let $H(t)$ be the location of the giraffe at time $t$. Obviously, $H(0) \in G^+(0)$ while $H(1) \notin G^+(1)$. The first instant $t_0$ when the giraffe leases the set $G^+(t)$ is defined by

$$t_0 = \inf_{0 \leq t \leq 1} \{t : H(t) \notin G^+(t)\}. \tag{3}$$

Since $G(t)$ is the boundary of $G^+(t)$, it is obvious that $H(t_0) \in G(t_0)$. Thus, at time $t_0$ the giraffe meets (at least) one of the horses.

*Proof of (b)*. Suppose that at time $t_0$, $0 < t_0 < 1$, the giraffe meets one of the horses. In order to meet another horse, he has to pass to another arc which can be done only through $O$ or through $A$, but since the velocity of the giraffe does not exceed unity (which is the velocity of the horses), then he cannot reach the point $A$ before $t = 1$, and if he moves through point $O$, then he would not reach another horse before $t = 1$. Q.E.D.

LEMMA 2. If the searcher uses the strategy $R(O, 0)$ (see Definition 1), then for any hiding strategy $H$

$$c(R(O, 0), H) \leq k. \tag{4}$$

*Proof*. Strategy $R(O, 0)$ for the searcher means: At time $t = 0$, the searcher makes an equiporbable choice from the horses of Lemma 1 and rides on the chosen horse to $A$; then at time $t = 1$, he makes an independent equiprobable choice from $k$ such horses which ride from $A$ to $O$, etc. We may look upon the hider as the giraffe, and thus Lemma 1(a) implies that if the searcher uses R(O, 0) and if the hider has not been captured until time $j - 1$, then the probability of capture in the period $t - 1 < t \leq j$ is at least $1/k$ (independent of the previous part of the trajectory). Thus,

$$\Pr(T > j) \leq \left(\frac{k-1}{k}\right)^{j}. \tag{5}$$

It follows that for any hiding trajectory H,

$$\begin{aligned} c(R(O, 0), H) &\leq \sum_{j=1}^{\infty} \Pr(j - 1 < T \leq j) \cdot j \\ &= \sum_{j=1}^{\infty} \Pr(T > j - 1) \\ &\leq \sum_{j=1}^{\infty} \left(\frac{k-1}{k}\right)^{j-1} = k. \qquad \text{(by (5))} \end{aligned}$$

Q.E.D.

LEMMA 3. Let RA be the strategy of the hider described as follows. Stay at point A until time $t = 1 - \varepsilon$ and then use $R(A, 1 - \varepsilon)$ (see Definition 1). Then for any search trajectory S

$$c(S, RA) \geq k - \varepsilon. \tag{6}$$

Proof. Use of the strategy $R(A, 1 - \varepsilon)$ means choosing each of the horses (note that it is the hider who is now riding the horses) with probability $1/k$ (and independently of previous choices) at the time instants $1 - \varepsilon, 2 - \varepsilon, \ldots$ . Now the searcher takes the role of the giraffe of Lemma 1. We shall use the fact that the velocity of the searcher does not exceed the velocity of the hider and Assumption 1 which states that the searcher cannot wait for the hider at any of the points O or A. Thus it can be assumed that at any one of the time instants $1 - \varepsilon$, $2 - \varepsilon, \ldots$, the probability that the searcher is either at O or at A is zero (this can be achieved by the searcher by using $\varepsilon$ as a random variable uniformly distributed in any small interval). Thus the condition of Lemma 1(b) is satisfied so that, for any trajectory of the searcher, if the hider has not been captured until $t = j - \varepsilon$, then the probability of capture in the time period $j - \varepsilon < t \leq j + 1 - \varepsilon$ is equal to $1/k$. Thus, the probability that capture will occur at the time period $j - \varepsilon < t \leq j + 1 - \varepsilon$ is equal to $((k - 1)/k)^{j-1}(1/k)$, hence

$$c(S, RA) \geq \sum_{j=1}^{\infty} \left(\frac{k - 1}{k}\right)^{j-1} \frac{1}{k}(j - \varepsilon) = k - \varepsilon. \qquad \text{Q.E.D.}$$

Theorem 1 is an immediate consequence of Lemmas 2 and 3. Actually, if $k > 1$, then $c(R(O, 0), H) < k$ for any hiding trajectory H. This is so because if $H(1) = A$, then $c(R(O, 0), H) = 1$, while if $H(1) \neq A$, then the capture time is equal to $\alpha$, $\alpha < 1$, with probability $1/k$. Thus

$$c(R(O, 0), H) \leq \frac{\alpha}{k} + \frac{k - 1}{k}(1 + k) < k.$$

Therefore, any mixed hiding strategy h satisfies $c(R(O, 0), h) < k$. Since $v = k$, it follows that the hider does not have an optimal strategy (but obviously has $\varepsilon$-optimal strategies), while the searcher does have an optimal strategy. This result is in accordance with the results presented in Appendix 1.

Remark 1. It follows from Lemmas 2 and 3 that the searcher can keep the probability of capture after t below $((k - 1)/k)^{[t]}$ ([t] is the integer part of t), while the hider can keep the probability of capture after t above $((k - 1)/k)^{[t+\varepsilon]}$. Thus, if instead of choosing the capture time $T$ as the cost function, we consider a more general cost function $U(T)$, where $U$ is any monotonic nondecreasing function, then the searcher can keep the value below

$$f(U) = \sum_{i=1}^{\infty} U(i) \frac{1}{k}\left(\frac{k-1}{k}\right)^{i-1} \tag{7}$$

while (for any $\varepsilon > 0$) the hider can keep the value above

$$f_{\varepsilon}(U) = \sum_{i=1}^{\infty} U(i - \varepsilon) \frac{1}{k}\left(\frac{k-1}{k}\right)^{i-1}$$

If $U$ is continuous on the left at the points $T = 1, 2, 3, \ldots$, then $f_{\varepsilon}(U) \to f(U)$ as $\varepsilon \to 0$.

In other words, if the cost is a continuous nondecreasing function of the capture time, then by using $R(0, 0)$, the searcher can guarantee the value $f(U)$, while the hider can guarantee $(1 - \varepsilon)f(U)$. It follows that in this case, the value of the game is $f(U)$ and the optimal ($\varepsilon$-optimal) strategies of the searcher and the hider are still the same as those described in Lemmas 2 and 3.

It is interesting to note that the value of the game is equal to $k$ irrespective of the maximal speed of the hider $w$, as long as it is not less than unity. If $w$ is less than unity, then the optimal strategies of the searcher and the hider may be quite complicated. However, using an argument similar to those which will be presented in Chapter 4, it seems to us that if $k$ is large and $w$ is not too small, then the value of the game should be approximately $k$, even for the case $0 < w < 1$, because the hider can achieve a value of $(1 - \varepsilon)k$ by randomly choosing one of the arcs and moving along it with speed w from A to O, then using an independent equiprobabilistic choice of another arc and moving along the chosen arc with maximal speed from O to A, and so on. If $m = 1/w$ and the searcher moves with maximal speed (unity), then, when reaching each of the points O or A, the maximal amount of information which may be available to the searcher is that at that moment the hider is not located on any of the m last arcs visited by the searcher. Thus, even if the searcher could rule out m arcs out of k each time he reaches O or A, then his gain would be to increase the probability of capture for each period $j < t \leq j + 1$ from $1/k$ to $1/(k - m) = 1/(k - (1/w))$. Thus the expected capture time would decrease at most to $k - (1/w)$, so that if k is large in comparison with $1/w$, the value obtained in Theorem 1 would remain about the same even for $w < 1$.

An interesting fact which is worth noting is the following. It has been shown in Chapter 2 that for an immobile hider, the value v of the search game on k parallel arcs is equal to $k/2$ for an even k and is approximately equal to $k/2$ for an odd k. Thus, for an immobile hider, $v \approx \mu/2g$. For a mobile hider, v is doubled and becomes $\mu/g$, and this is due to the fact that, contrary to the case of an immobile hider, the searcher cannot rule out the arcs previously visited by him. We shall have the same phenomenon for the case of a two-dimensional search space.

## 3.3 SEARCH ON A CIRCLE

In this section, we consider searching for a mobile hider on a circle. As in Section 3.2, we shall make the assumption that the maximal velocity w of the hider is greater than or equal to 1. By denoting the (known) starting point of the searcher by O and its antipode by A, the problem reduces to a special case of the search game solved in Section 3.2 with $k = 2$ so that the value of the game is the length of the circle $\mu$. However, we must be aware of the fact that in Section 3.2 the strategies of the searcher were restricted by Assumption 1, which require that the searcher cannot wait at O or at A. We now show that the result for $k = 2$ remains valid even without this assumption. The searcher's strategy $R(O, 0)$ (see Definition 1) which guarantees an expected capture time $\mu$ (= 2) remains the same, but the hider's strategy needs a little modification. At time $t = 0$, the hider should choose a random variable $\theta$ which has a uniform distribution in $(0, \varepsilon)$, $\varepsilon$ being a small number, and then choose a point $O'$ as one of the points which has a distance $\theta$ from O. The antipode of $O'$ is denoted by $A'$. It is easy to see that by staying at $A'$ until $t = 1 - \varepsilon$ and then using $R(A', 1 - \varepsilon)$, i.e., moving from $A'$ to $O'$ randomly, the hider can make sure that the capture time will exceed $2 - \varepsilon$ $(= \mu - \varepsilon)$, because the probability of capture at $O'$ or at $A'$ is zero, so that the argument used in Lemma 3 is valid here as well.

It should be noted that the result $v = \mu$ depends on the assumption that the initial starting point of the searcher is known $(S(0) = O)$. This result is no longer valid if one changes this assumption. For example, if one assumes that $S(0)$ has a uniform distribution on the circle, as was done by Alpern (1974), then the value is $3\mu/4$. This result is demonstrated by the following theorem which is taken from Alpern (1974).

THEOREM 2. Consider the search game with a mobile hider on the circle, assuming that $S(0)$ and $H(0)$ are uniformly and independently distributed on the circle. Then

$$v = 3\mu/4. \tag{8}$$

An optimal strategy for each player is to oscillate at speed 1 between his initial point and its antipode, each time making an equiprobable choice between the clockwise and counterclockwise directions.

Proof. Assume for convenience that $\mu = 2$. For any $t > 0$, let $I = [t]$ and $\alpha = t - [t]$. Let

$$L(t) = 1/2^I - \alpha/2^{I+1}. \tag{9}$$

At first we show that under the searching strategy $s^*$, described by Theorem 2, the probability of capture after time t satisfies

$$\Pr(T > t) \leq L(t) \tag{10}$$

for any hiding trajectory H.

We prove (10) as follows. Assume without loss of generality that I is even. Denote

$$\left.\begin{array}{ll} E, & \text{the event: } d(O, H(t)) \leq \alpha, \\ \bar{E}, & \text{the complement of } E. \end{array}\right\} \tag{11}$$

Then the same considerations used in Lemmas 1 and 2 lead to the inequalities

$$\Pr(T > t|E) \leq 1/2^{I+1} \tag{12}$$

and

$$\Pr(T > t|\bar{E}) \leq 1/2^I. \tag{13}$$

Since $O = S(0)$ is uniformly distributed on the circle, it follows from (11) that

$$\Pr(E) = \alpha, \tag{14}$$

and (10) readily follows from (12) to (14).

Now,

$$c(s^*, H) = \int_0^\infty \Pr(T > t)\, dt = \sum_{I=0}^{\infty} \int_0^1 \Pr(T > I + \alpha)\, d\alpha$$

$$\leq 3/2 \qquad \text{(by (9) and (10))}. \tag{15}$$

The optimality of the hider's strategy $h^*$, described by Theorem 2, is established in a similar manner by proving that for any trajectory S, if the hider uses $h^*$, then

$$\Pr(T > t) = L(t), \tag{16}$$

and thus $c(S, h^*) = 3/2$. Q.E.D.

Note that the optimal search strategy is the same as in the case that S(0) is a fixed known point, while the hiding strategy has to be modified due to the fact that the hider does not know S(0).

The value would still be $3\mu/4$ in the case that both S(0) and H(0) can be chosen by the players because each one of them can gaurantee $3\mu/4$ by choosing a uniformly distributed starting point and then using the strategy described by Theorem 2.

In contrast to the foregoing discussion, we shall see in Chapter 4 that in the case of multidimensional search, the value of the game is not sensitive to any assumptions about the location of S(0).

## 3.4 AN OPEN PROBLEM

In Chapter 2, where search games with an immobile hider were considered, we proved Theorem 2 which stated that the value of the search game lies between $\mu/2$ and $\mu$, where $\mu$ is the sum of the lengths of the arcs of the graph Q. The lower bound is achieved in the case that Q is an Eulerian graph, while the upper bound is achieved for the case where Q is a tree.

It seems to us that analogous results should hold for the case of a mobile hider. Our conjecture is that under our usual assumption that S(0) is a fixed point known to the hider, the value of the game should be bounded below by $\mu$ and above by $2\mu$. The case considered in Section 3.2, where Q consists of k parallel arcs, is an example in which $v = \mu$. A case where $v \approx 2\mu$ could be of the following type.

Let Q consist of k rays radiating from the origin O, each ray of equal length 1. For this case it is obvious that $v < 2\mu$, because if at time $t = 0, 2, 4, \ldots,$ the searcher chooses each ray with probability $1/k$ and then moves along it and returns to the origin, then the expected capture time is at most

$$\sum_{j=1}^{\infty} (1 + 2(j - 1))\left(\frac{k - 1}{k}\right)^{j-1} \frac{1}{k} < 2k = 2\mu.$$

On the other hand, if k is large, then the hider can achieve a capture time exceeding $(1 - \varepsilon)2k$ by a strategy which is similar to the one which will be presented in Section 4.3: He uniformly chooses a ray and stays at its terminal point until time J, where $1 << J << k$; then he makes another independent equiprobable choice among the rays, moves to the terminal point of the chosen ray, with maximal speed, and stays there until time 2J, etc. (It should be assumed as in Section 3.2 that the searcher cannot wait at the origin O or alternatively that the hider can pass from each ray to another by moving "near" the origin without actually passing through it.)

We conjecture that the two preceding cases represent the upper and the lower bounds for the value of the search game with a mobile hider (with $w \geq 1$) in a graph, i.e., that the inequality $\mu \leq v \leq 2\mu$ holds for any graph.

For the lower bound to hold, one probably needs an assumption similar to Assumption 1, namely, that the searcher cannot stay at vertices of Q. It should also be noted that if we allow the searcher an arbitrary starting point, then the lower bound for the value decreases to $\mu/2$, because $v = \mu/2$ in the case that Q is a line segment [a, b] and the searcher can choose his starting point. (The searcher can guarantee an expected capture time not exceeding $(b - a)/2$ by choosing $S(0) = a$ with probability 1/2 and $S(0) = b$ with probability 1/2, and moving with maximal speed to the other end of the segment, while the hider can guarantee an expected capture time greater or equal to $(b - a)/2$ by choosing a point H which is uniformly distributed in [a, b], and staying at H.)

It also may be interesting to investigate the effect of removing Assumption 1 on the optimal solution of search games with a mobile hider in a graph (e.g., the search on k arcs solved in Section 3.2).

Chapter 4

# Mobile Hider In A Multidimensional Region

## 4.1 GENERAL DESCRIPTION

In this chapter, the domain of search is a two- (or more) dimensional region in a Euclidean space. The searcher can move along any continuous trajectory which starts from the origin O. The hider can choose his initial location and an arbitrary continuous trajectory starting from that point, and can move along his trajectory with maximal speed w. In contrast to Chapter 3, we shall not require that $w \geq 1$, but we shall assume that w is not too small (the exact formulation of this condition will be presented in Section 4.3).

The notations that will frequently be used in this chapter are: $\mu$ is the Lebesgue measure of Q, D the diameter of Q, and g the maximal rate of discovery of the searcher (which is equal to 2r for two-dimensional sets, $\pi r^2$ for three-dimensional sets, etc.). The radius of detection r will be assumed to be small in relation to the magnitude of Q. We shall proceed under the assumption that the detection radius is a constant and the maximal velocity of the searcher is 1, but in Section 4.4 we show how to extend the results to the case in which both the radius of detection and the maximal velocity of the searcher depend on the location inside Q.

For the case of an immobile hider in a two-dimensional region, it has been shown in Section 2.5 that the value of the search game, v, satisfies $v \sim \mu/2g$.

For the case of a mobile hider, we shall show in Section 4.2 that the searcher can guarantee an expected capture time not exceeding $(\mu/g)(1 + \varepsilon)$. The dual result is presented in Section 4.3. We show there that if Q is covex, then the hider can make sure that the expected capture time will exceed $(\mu/g)(1 - \varepsilon)$. Thus, we demonstrate that the value of the princess and monster game in a multidimensional convex region satisfies $v \sim \mu/g$, independently of the shape of the search space, and we present $\varepsilon$-optimal strategies for both players. We conduct the detailed proofs for two-dimensional regions, but in Section 4.4 we show how to extend the results to higher-dimensional regions, and also discuss some other extensions. In Section 4.5, we show that the $\varepsilon$-optimal strategies are still $\varepsilon$-optimal even if we use a more general cost function. We conclude this chapter by presenting some interesting open problems in Section 4.6.

## 4.2 STRATEGY OF THE SEARCHER

In this section, we shall prove that for any bounded two-dimensional region Q which satisfies a rather weak requirement, if the detection radius r is small, then the searcher has a strategy $s^*$ which makes sure that the expected capture time does not exceed $(1 + \varepsilon)(\mu/2r)$, where $\varepsilon$ is small. We shall

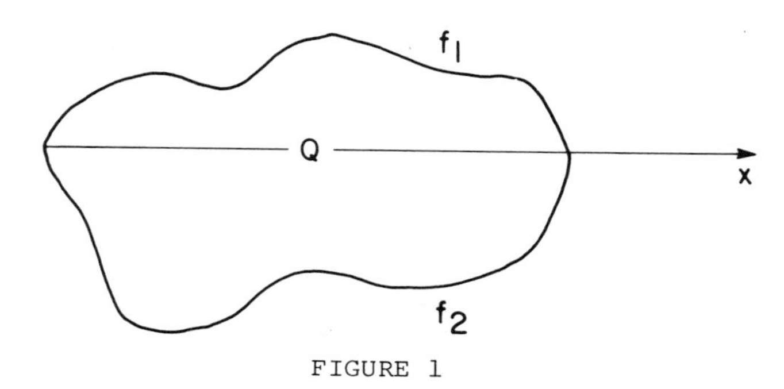


FIGURE 1

carry the details of the proof for a simple region Q which satisfies:

$$\left.\begin{array}{l}\text{Q is a compact region whose boundary is the}\\ \text{union of two single-valued continuous}\\ \text{functions } f_1 \text{ and } f_2 \text{ as depicted in Figure 1,}\end{array}\right\} \quad (1)$$

but the same proof is also applicable for any region which satisfies Assumption 1 of Section 2.5 (i.e., Q is a disjoint union of a finite number of simple regions).

The search strategy $s^*$ to be considered has the following general structure. Q is covered by a set of parallel and similar narrow rectangles $Q_1,\ldots,Q_m,\ldots,Q_M$; a rectangle $Q_m$ is randomly chosen and entered; $Q_m$ is examined by moving N times forward and backward along randomly chosen trajectories in $Q_m$; then another rectangle is randomly chosen, etc. The number N should be large enough to "absorb" the effect of the time spent in going from one rectangle to another, but on the other hand, N must not be so large that too much time is spent in one rectangle. Having this idea in mind we proceed as follows.

Referring to Assumption 1 and Figure 1, let $Q_a \supset Q$ be the set with minimal area which is a union of the rectangles

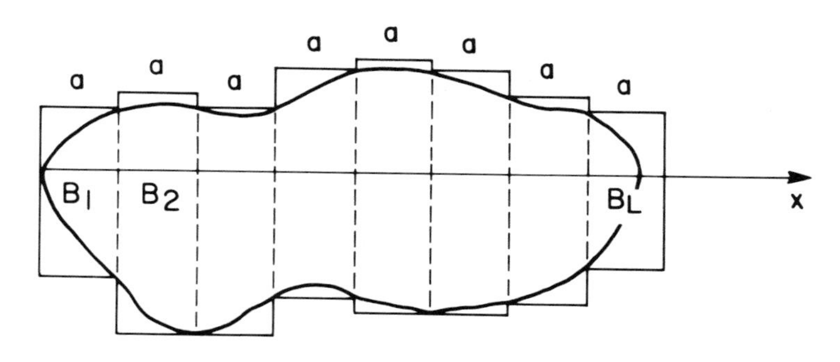


FIGURE 2

$B_1, \ldots, B_L$, where all these rectangles have the same width a and are parallel to the x axis as shown in Figure 2.

Let $\mu$ be the area of Q and $\mu_a$ be the area of $Q_a$. Then

$$\mu_a = (1 + \phi_a)\mu, \tag{2}$$

where, by Assumption 1,

$$\phi_a \downarrow 0 \quad \text{as} \quad a \downarrow 0. \tag{3}$$

For any $Z_1$, $Z_2 \in Q_a$, let $d(Z_1, Z_2)$ be the minimal length of a path which connects $Z_1$ and $Z_2$ and passes inside $Q_a$. We define the diameter D of $Q_a$ as

$$D = \max_{Z_1, Z_2 \in Q_a} d(Z_1, Z_2). \tag{4}$$

We shall give a constructive proof of the following theorem.

THEOREM 1. Let r satisfy

$$r = \varepsilon_a(a^2/2D) \tag{5}$$

and assume that

$$\delta = \varepsilon_a^{1/4} << 1. \tag{6}$$

Then there exists a search strategy $s^*$ in $Q_a$ such that for any evading trajectory H used by the hider, the expected capture

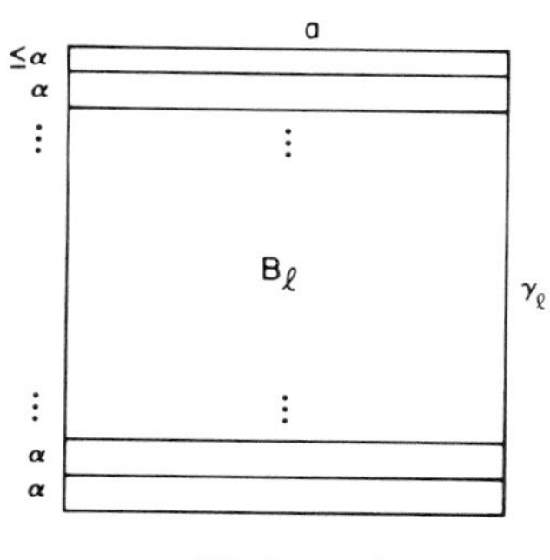


FIGURE 3

time $c(s^*, H)$ satisfies

$$c(s^*, H) \le (1 + 4\delta)(\mu_a/2r) = (1 + 4\delta)(1 + \phi_a)(\mu/2r). \tag{7}$$

Proof. Let

$$\alpha = \delta^2 a. \tag{8}$$

Divide each rectangle $B_l$, $l = 1,\ldots, L$ (see Figure 2), into "narrow" rectangles so that all of these rectangles, except possibly one, have a width $\alpha$, while the upper one has a width $\le\alpha$, as depicted in Figure 3. Thus, each rectangle $Q_m$, $m = 1,\ldots, M$, has a width $\le\alpha$, and the number M of such rectangles satisfies

$$M = \sum_{l=1}^{L} \left[\left(\frac{\gamma_l}{\alpha}\right) + 1\right] \le \sum_{l=1}^{L} \frac{\gamma_l}{\alpha} + L$$

$$\le \frac{\mu_a}{a\alpha} + \frac{D + a}{a} = \frac{\mu_a}{a\alpha}\left(1 + \alpha\frac{D + a}{\mu_a}\right)$$

$$= \frac{\mu_a}{a\alpha}\left(1 + \delta^2 \frac{a(D + a)}{\mu_a}\right) < \frac{\mu_a}{a\alpha}\left(1 + \frac{\delta}{2}\right) \tag{9}$$

(by (8) and (6)).

Let N be a positive integer defined by

$$N = [D/\delta a] + 1. \tag{10}$$

Let $\alpha'$ be a real number. If $\alpha' \ge 2r$, then we define a random variable y with the density

$$
\left.\begin{aligned}
f(y) &= \frac{2}{\alpha' + 2r} && \text{for} \quad 0 \le y \le r, \\
&= \frac{1}{\alpha' + 2r} && \text{for} \quad r < y < \alpha' - r, \\
&= \frac{2}{\alpha' + 2r} && \text{for} \quad \alpha' - r \le y \le \alpha',
\end{aligned}\right\} \qquad (11)
$$

$$
= 0 \qquad \text{elsewhere.}
$$

If $\alpha' < 2r$, then we define y to be identically $\alpha'/2$.

The search strategy $s^*$ is composed of independent repetitions of the following step. At time $t = 0$ make a random choice out of the narrow rectangles $Q_1, Q_2, \ldots, Q_m, \ldots, Q_M$ such that each rectangle $Q_m$ has a probability $1/M$ of being chosen, and also make a random choice of N independent random variables $y_1, \ldots, y_N$, where N is given by (10) and all the $y_n$, $n = 1, \ldots, N$, have the probability density given by (11), where $\alpha'$ is the width of the chosen rectangle. In order to describe the motion of the searcher within $Q_m$, we shall use a coordinate system with origin at the lower left corner of $Q_m$ as depicted in Figure 4. At time $t = 0$, the searcher starts moving as fast as possible to the point $(0, y_1)$. He rests at that point until $t = D$, and then moves with maximal velocity in a straight line to the point $(a, y_1)$ and reaches it at time $t = D + a$; then he moves to the point $(a, y_2)$ and rests there until $t = D + a + \alpha$, and at that moment he starts moving to $(0, y_2)$, etc. The important feature of the movement of the

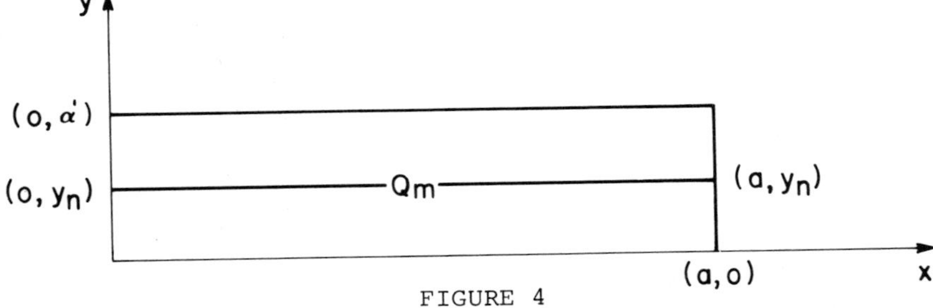


FIGURE 4

searcher is that at the time segments $\tau_n = [D + (n - 1)(a + \alpha), D + (n - 1)(a + \alpha) + a]$, $n = 1, \ldots, N$, he moves along the intervals which join $(0, y_n)$ to $(a, y_n)$. We shall show that for this kind of movement, the following lemma holds.

LEMMA 1. If the searcher moves in the manner already described, then the probability p of capture during the time segment $0 < t \leq D + N(a + \alpha)$ satisfies

$$p \geq \frac{1}{M}\left(1 - \left(\frac{\alpha}{\alpha + 2r}\right)^N\right). \tag{12}$$

Proof. Consider a specific time segment $\tau_n$ given by

$$\tau_n = \{t : D + (n - 1)(a + \alpha) \leq t \leq D + (n - 1)(a + \alpha) + a\}, \tag{13}$$

where $1 \leq n \leq N$.

We shall distinguish between two cases. If n is odd, then for any $t \in \tau_n$, we define $G_m(t)$, $m = 1, 2, \ldots, M$, as the vertical line segment of width $\alpha'$, which has a distance $d(t)$ from the left vertical edge of $Q_m$, where

$$d(t) = t - (D + (n - 1)(a + \alpha)), \tag{14}$$

so that $G_m(t)$ is given by

$$G_m(t) = \{(d(t), y), 0 \leq y \leq \alpha'\}. \tag{15}$$

If n is even, then $G_m(t)$ is the vertical line segment which has the distance $d(t)$ given by (14) from the right vertical edge of $Q_m$; i.e., in this case

$$G_m(t) = \{(a - d(t), y), 0 \leq y \leq \alpha'\}. \tag{16}$$

In both cases we define

$$G(t) = \bigcup_{m=1}^{M} G_m(t). \tag{17}$$

By an argument similar to those used in proving Lemma 1 of Section 3.2, one can show that the following proposition holds.

PROPOSITION. If H is any trajectory used by the hider, then for any n there exists at least one time instant $t_n \in \tau_n$ (see (13)) such that the point $H(t_n)$ visited by the hider at time $t_n$ satisfies $H(t_n) \in G(t_n)$ (see (17)).

It follows from the proposition that for any n there exists an m such that

$$H(t_n) \in G_m(t_n). \tag{18}$$

Let $I_m(n) = 1$ if (18) holds and zero otherwise, and define

$$I_m = \sum_{n=1}^{N} I_m(n). \tag{19}$$

Then it follows from the foregoing discussion that

$$\sum_{m=1}^{M} I_m \geq N. \tag{20}$$

If the searcher chooses the rectangle $Q_m$ and if $I_m(n) = 1$, then it follows from the definition of the random variable $y_n$ that the probability of capture during the time segment $\tau_n$ is greater than or equal to the probability that at the time $t_n$ (see (18)) the random interval $Y_n$ given by

$$Y_n = [y_n - r, y_n + r] \cap [0, \alpha']$$

will contain the y-th coordinate of $H(t_n)$. Now, it follows from (11) that for any point b in the interval $[0, \alpha']$ the probability that $b \in Y_n$ is greater than or equal to $2r/(\alpha' + 2r) \geq 2r/(\alpha + 2r)$.

Since the random variables $y_1, \ldots, y_N$ are independent, it follows that if $Q_m$ has been chosen, then the probability of capture during the time segment $0 < t \le D + N(a + \alpha)$ is greater than or equal to

$$1 - \left(1 - \frac{2r}{\alpha + 2r}\right)^{I_m} = 1 - \left(\frac{\alpha}{\alpha + 2r}\right)^{I_m} \qquad \text{(see (19))}.$$

Since each rectangle $Q_m$, $m = 1, \ldots, M$, is chosen with probability $1/M$, if follows that the probability $p$ of capture during the time segment $0 < t \le D + N(a + \alpha)$ satisfies

$$p \ge \sum_{m=1}^{M} \frac{1}{M}\left(1 - \left(\frac{\alpha}{\alpha + 2r}\right)^{I_m}\right)$$

$$= 1 - \frac{1}{M} \sum_{m=1}^{M} \left(\frac{\alpha}{\alpha + 2r}\right)^{I_m} \tag{21}$$

Since for any nonnegative integers $J$, $K$,

$$\left(1 - \left(\frac{\alpha}{\alpha + 2r}\right)^{J}\right)\left(1 - \left(\frac{\alpha}{\alpha + 2r}\right)^{K}\right) \ge 0,$$

it follows that

$$\left(\frac{\alpha}{\alpha + 2r}\right)^{J} + \left(\frac{\alpha}{\alpha + 2r}\right)^{K} \le 1 + \left(\frac{\alpha}{\alpha + 2r}\right)^{J+K}$$

so that

$$\sum_{m=1}^{M} \left(\frac{\alpha}{\alpha + 2r}\right)^{I_m} \le M - 1 + \left(\frac{\alpha}{\alpha + 2r}\right)^{\sum_{m=1}^{M} I_m}$$

$$\le M - 1 + \left(\frac{\alpha}{\alpha + 2r}\right)^{N} \tag{22}$$

(by (20)).

Thus, it follows from (21) and (22) that

$$p \ge \frac{1}{M}\left(1 - \left(\frac{\alpha}{\alpha + 2r}\right)^{N}\right),$$

so that the proof of Lemma 1 is completed.

We now proceed with the proof of Theorem 1. First we note that (5), (6), (8), and (10) imply that

$$\frac{2r}{\alpha} = \frac{2\delta^4 a^2}{2D\delta^2 a} = \delta^2 \frac{a}{D} \geq \frac{\delta}{N}.$$

Thus by (12), the probability p of capture in the time segment $0 < t \leq D + N(a + \alpha)$ satisfies

$$p \geq \frac{1}{M}\left(1 - \left(\frac{1}{1 + 2r/\alpha}\right)^N\right) \geq \frac{1}{M}\left(1 - \frac{1}{(1 + \delta/N)^N}\right)$$

$$\geq \frac{1}{M}\left(1 - \frac{1}{1 + \delta}\right) = \frac{\delta}{M(1 + \delta)} \tag{23}$$

Now, since the search strategy $s^*$ is composed of independent repetitions of the step described for the time segment $0 \leq t \leq D + N(a + \alpha)$, then for any hiding trajectory H, the probability $\overline{p}_K$ of capture after the time instant $t = K(D + N(a + \alpha))$ satisfies

$$\overline{p}_K \leq (1 - p)^K. \tag{24}$$

Thus the expected capture time $c(s^*, H)$ satisfies (see (5) of Chapter 1)

$$c(s^*, H) \leq (D + N(a + \alpha)) \sum_{K=0}^{\infty} \overline{p}_K \leq \frac{D + N(a + \alpha)}{p}$$

(by (24))

$$\leq \frac{M(1 + \delta)}{\delta} Na\left(1 + \frac{D}{Na} + \frac{\alpha}{a}\right) \qquad \text{(by (23))}$$

$$\leq \frac{\mu_a}{a\alpha}\left(1 + \frac{\delta}{2}\right) \frac{1 + \delta}{\delta}\left(\frac{1}{\delta}\frac{D}{a} + 1\right) a(1 + \delta + \delta^2)$$

(by (8)-(10))

$$= \frac{1}{\delta^4} \frac{\mu_a D}{a^2}\left(1 + \frac{\delta}{2}\right)(1 + \delta)\left(1 + \delta\frac{a}{D}\right)(1 + \delta + \delta^2)$$

$$\leq \frac{\mu_a}{\varepsilon_a(a^2/D)} (1 + 4\delta) = \frac{\mu_a}{2r} (1 + 4\delta) \tag{25}$$

(by (5) and (6)) Q.E.D.

The following corollary will enable us to establish, in Section 4.5, the $\varepsilon$-optimality of $s^*$ for a more general cost function.

COROLLARY 1. If the searcher uses the strategy $s^*$ used in the proof of Theorem 1, then for any hiding trajectory H the probability that the capture time T exceeds t satisfies

$$\Pr(T > t) \leq (1 + \varepsilon) \exp(-2rt/(1 + \varepsilon)\mu)$$

with $\varepsilon \to 0$ as $r \to 0$.

<u>Proof</u>. Let

$$I_t = \left[\frac{t}{D + N(a + \alpha)}\right].$$

Then it follows from (23) and (25) that if the searcher uses $s^*$, then for any H

$$\Pr(T > t) \leq \Pr(T > I_t(D + N(a + \alpha)))$$

$$= (1 - p)^{I_t} \leq \left(1 - \frac{\delta}{M(1 + \delta)}\right)^{I_t} \leq \exp(-I_t\delta/M(1 + \delta))$$

$$\leq \exp\left(-\frac{t\delta}{(1 + \delta)M(D + N(a + \alpha))} + \frac{\delta}{M(1 + \delta)}\right)$$

by (2), (5), (6), and (8)-(10):

$$\leq \exp\left(-\frac{2rt}{\mu(1 + \phi_a)(1 + 4\delta)} + \frac{\delta}{M(1 + \delta)}\right).$$

Since both $\delta \to 0$ and $\phi_a \to 0$ as $r \to 0$, we obtain the desired result.

It should be noted that if the searcher uses the strategy $s^*$ presented in the proof, then a part of his trajectory, which is near the boundary of $Q_a$, might be slightly outside of the

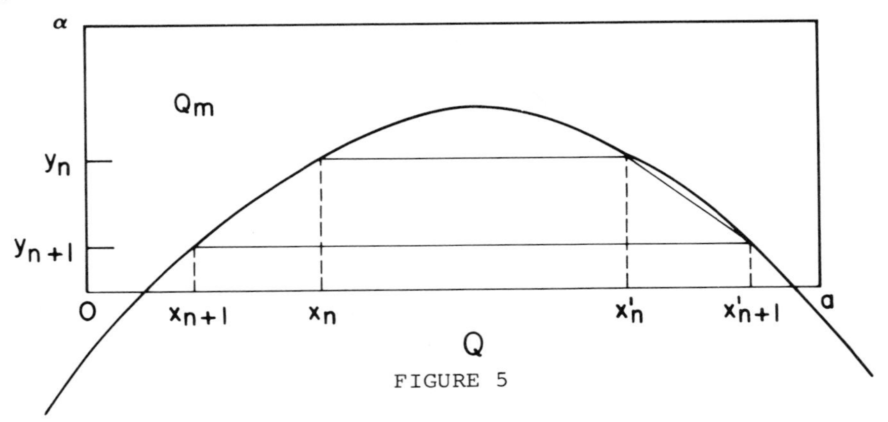


FIGURE 5

original search space Q. However, if Q is convex, then we can make a slight modification in $s^*$ and introduce a search strategy $s^{**}$ which uses trajectories entirely inside Q and still guarantees that the result (7) of Theorem 1, holds. The strategy $s^{**}$ is defined as follows. If the chosen rectangle $Q_m$ is inside Q, then the movement is identical to the one in $s^*$. However, if a part of $Q_m$ is outside Q, as depicted in Figure 5, we make the following modification.

Assume that in the strategy $s^*$ the searcher moves from the point $(0, y_n)$ to $(a, y_n)$ in the time segment $\tau_n$ (see (13)) and then moves from the point $(a, y_{n+1})$ to $(0, y_{n+1})$ in the time segment $\tau_{n+1}$. The movement in $s^{**}$ is as follows. The searcher moves from the point $(x_n, y_n)$ to $(x_n', y_n)$ (see Figure 5) in the time segment $D + (n - 1)(a + \alpha) + x_n \leq t \leq D + (n - 1)(a + \alpha) + x_n'$, then moves with maximal velocity in a straight line from $(x_n', y_n)$ to $(x_{n+1}', y_{n+1})$ and stays there until the time instant $t' = D + n(a + \alpha) + a - x_{n+1}'$ (the searcher can arrive at $(x_{n+1}', y_{n+1})$ before $t'$ because the length of the segment which joins $(x_n', y_n)$ to $(x_{n+1}', y_{n+1})$ does not exceed $2a + \alpha - x_n' - x_{n+1}'$), then moves from $(x_{n+1}', y_{n+1})$ to $(x_{n+1}, y_{n+1})$ in the time segment $t' \leq t \leq D + n(a + \alpha) + a - x_n$, etc.

By using exactly the same method of proving (7), it can be established that for any hiding trajectory H, the expected capture time satisfies

$$c(s^{**}, H) \leq (1 + 4\delta)(\mu_a/2r),$$

so that for a convex Q, the strategy $s^{**}$ guarantees the desired result by moving only inside Q.

## 4.3 STRATEGY OF THE HIDER

The strategy $h_u$ of the hider which is considered in this section is defined as follows.

DEFINITION 1. Choose a point $Z_1$, using a uniform probability distribution in Q, and stay there during the time period $0 \leq t < u$. At the time $t = u$, choose a point $Z_2$, which is uniformly distributed in Q independent of $Z_1$, move toward $Z_2$ with velocity $w_1 = \min(w, 1)$ in a straight line, and stay in $Z_2$ for a time period of length u; then choose a point $Z_3$ uniformly distributed in Q and independent of $Z_1$ and $Z_2$, move toward it, again with velocity $w_1 = \min(w, 1)$ in a straight line, and stay there for a time period of length u, and so on.

The "resting time" u should satisfy two conditions:

(1) It should not be too long, so that the area covered by the searcher in a time interval of length u would be small relatively to $\mu$; but on the other hand:

(2) In order to keep the probability of capture during motion relatively small, the hider should not move too frequently and thus u should not be too short.

Assume that the hider uses a strategy $h_u$ as described by

Definition 1. Let

$$E_i, \text{ the event: capture occurs at point } Z_i, \quad i = 1, 2, \ldots . \tag{26}$$

Now if it were possible to neglect the probability of capture during the motion from $Z_i$ to $Z_{i+1}$, then for any search trajectory S, the expected capture time would approximately satisfy

$$c(S, h_u) \gtrsim \sum_{i=1}^{\infty} u\,(i-1)\Pr(E_i)$$

$$= u \sum_{n=1}^{\infty} \sum_{i=n+1}^{\infty} \Pr(E_i) \geq u \sum_{n=1}^{\infty} \Pr\left(\overline{\bigcup_{i=1}^{n} E_i}\right). \tag{27}$$

Since each $Z_i$ is uniformly distributed in Q and since the maximal rate of discovery is 2r, it then follows that for each $Z_i$, the probability of being discovered at $Z_i$ is at most $2ru/\mu$. Thus, it follows from the independence of $Z_i$ that

$$\Pr\left(\overline{\bigcup_{i=1}^{n} E_i}\right) \geq \left(1 - \frac{2ru}{\mu}\right)^n. \tag{28}$$

Thus, it would follow from (27) and (28) that for all S

$$c(S, h_u) \gtrsim u \sum_{i=1}^{n} \Pr\left(\overline{\bigcup_{i=1}^{n} E_i}\right) \geq u\frac{1 - (2ru/\mu)}{2ru/\mu} = \frac{\mu}{2r} - u, \tag{29}$$

so that if the parameter u of the hider's strategy $h_u$ would be chosen to be small in comparison with $\mu/2r$, we would get the desired result.

We have presented the previous discussion in order to help the reader understand the motivation behind the definition of the strategy $h_u$ and the idea of the (rather complicated) proof of Theorem 2 which follows.

We must also require that the maximal velocity of the hider

w should not be too small, because when w reduces to zero we approach the situation of an immobile hider considered in Chapter 2, and the value of the search game should then be approximately $\mu/2g$ $(= \mu/4r)$. Such a condition is the following:

$$(rD/\mu)\max(1/w, 1) < \varepsilon' \ll 1. \tag{30}$$

In the next theorem we show for any two-dimensional convex set, that if (30) holds, then the hider can make sure that the capture time will exceed $(1 - \varepsilon)\mu/2r$. The theorem is formulated and proved for two dimensions but it can be easily extended to three or more dimensions.

THEOREM 2. Let the search space Q be any two-dimensional convex set, and let condition (30) be satisfied. Denote

$$\delta = (\varepsilon'\cdot 37\pi \max(D^2/\mu, 1))^{1/4} \tag{31}$$

and assume that $\delta \ll 1$.

If the hider uses the strategy $h_u$ presented in Definition 1, where

$$u = \delta(\mu/2r), \tag{32}$$

then for any search trajectory S, the expected capture time satisfies

$$c(S, h_u) \geq (\mu/2r)(1 - 3\delta). \tag{33}$$

Proof. Let $Z_i$, $i = 1, 2, \ldots$, be the hiding points of Definition 1 and let

$$\left.\begin{array}{lll} F_i, & \text{the event:} & \text{capture occurs while the hider is moving from point } Z_i \text{ to point } Z_{i+1},\ i = 1, 2, \ldots, \\ F_i', & \text{the event:} & \text{capture does not occur before the hider leaves point } Z_i. \end{array}\right\} \tag{34}$$

Then (see (26) and (34))

$$\Pr(F_n') = \Pr\left(\overline{\bigcup_{i=1}^{n} E_i \bigcup_{i=1}^{n-1} F_i}\right) = 1 - \Pr\left(\bigcup_{i=1}^{n} E_i \bigcup_{i=1}^{n-1} F_i\right)$$

$$\geq 1 - \Pr\left(\bigcup_{i=1}^{n} E_i\right) - \Pr\left(\bigcup_{i=1}^{n-1} F_i\right)$$

$$\geq \Pr\left(\overline{\bigcup_{i=1}^{n} E_i}\right) - \sum_{i=1}^{n-1} \Pr(F_i). \tag{35}$$

The first stage of the proof is to establish an upper bound for $\Pr(F_i)$ so as to show that it is negligible in comparison to the other relevant terms.

Since the hider moves from point $Z_i$ to point $Z_{i+1}$ in a straight line with velocity $w_1 = \min(w, 1)$, it follows that the time of movement $\tau_i$ satisfies

$$\tau_i \leq D/w_1 = D/\min(w, 1). \tag{36}$$

Let

$$J_i = [\tau_i/r] + 1 \leq (D/w_1 r) + 1, \tag{37}$$

and let $S_i$ be that section of the searcher's trajectory which corresponds to a time interval of length $rJ_i$, starting from the moment when the hider leaves $Z_i$. Divide $S_i$ into $J_i$ arcs $S_{i1}, \ldots, S_{iJ_i}$ such that

each arc $S_{ij}$ corresponds to an equal time interval of length r. (38)

Let $F_{ij}$ be the event:

$F_{ij}$: at some point of $S_{ij}$, the distance between the searcher and the hider is less than or equal to r. (39)

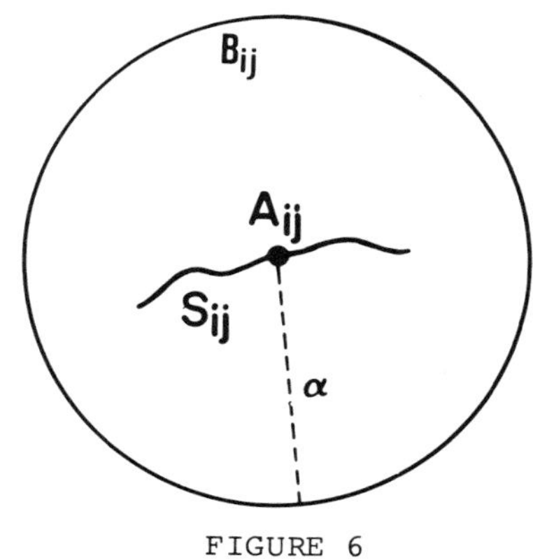


FIGURE 6

Obviously (see (34)),

$$\Pr(F_i) \le \Pr\left(\bigcup_{j=1}^{J_i} F_{ij}\right) \le \sum_{j=1}^{J_i} \Pr(F_{ij}). \tag{40}$$

Let us denote the middle point of the arc $S_{ij}$ by $A_{ij}$, the time when the searcher reaches this point by $t_{ij}$, and let $B_{ij}$ be a circle of radius $\alpha$ around $A_{ij}$ (as depicted in Figure 6), where $\alpha$ satisfies

$$\alpha = \left(1 + \frac{1}{2} + \frac{w_1}{2}\right)r. \tag{41}$$

It follows from (38) that a necessary condition for the validity of the event $F_{ij}$ (see (39)) is the event $M_{ij}$:

$M_{ij}$, the event: at the time $t_{ij} = (j + 1/2)r$ (corresponding to the point $A_{ij}$) the hider is inside the circle $B_{ij}$. (42)

and a necessary condition for the validity of $M_{ij}$ is the event $D_{ij} \cap D^*_{ij}$, where

$D_{ij}$, the event: the distance $d_{ij}$ of $Z_i$ from $A_{ij}$ satisfies $(j + 1/2)w_1 r - \alpha \le d_{ij}$ $\le (j + 1/2)w_1 r + \alpha$ (see (41)) (43)

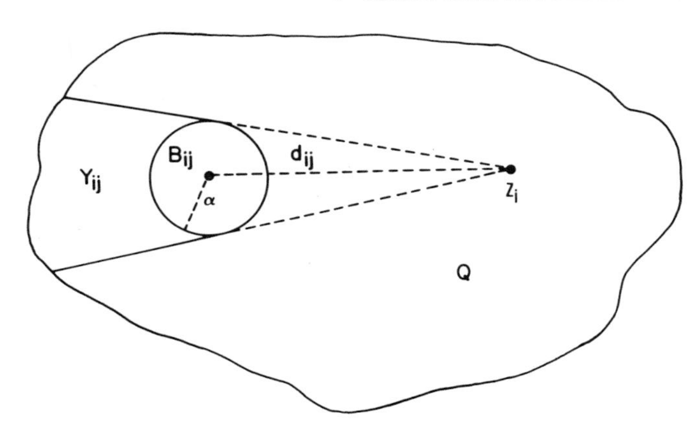


FIGURE 7

and

$$D^*_{ij}, \text{ the event: } Z_{i+1} \text{ lies in the region } Y_{ij} \text{ depicted in Figure 7.} \tag{44}$$

Obviously,

$$\Pr(D_{ij}) \leq \frac{\pi(2j+1)w_1 r\cdot 2\alpha}{\mu} = \frac{\pi(2j+1)w_1(3+w_1)r^2}{\mu} \quad \text{(by (41))} \tag{45}$$

and

$$\Pr(D^*_{ij}|D_{ij}) \leq \frac{D}{\mu}\,\frac{2\alpha D}{(j+1/2)w_1 r - \alpha} = \frac{(3+w_1)D^2}{(jw_1 - 3/2)\mu}. \tag{46}$$

We are now in a position to obtain an upper bound for $\Pr(F_i)$. Let

$$m = \min([\sqrt{D/w_1 r}] - 1,\ J_i). \tag{47}$$

Then

$$\begin{aligned}\Pr(F_i) &\leq \sum_{j=1}^{m} \Pr(F_{ij}) + \sum_{j=m+1}^{J_i} \Pr(F_{ij}) \\ &\leq \sum_{j=1}^{m} \Pr(D_{ij}) + \sum_{j=m+1}^{J_i} \Pr(D_{ij}) \cdot \Pr(D^*_{ij}|D_{ij}).\end{aligned} \tag{48}$$

Now

$$\sum_{j=1}^{m} \Pr(D_{ij}) \le \sum_{j=1}^{m} \frac{\pi(2j + 1)w_1(3 + w_1)r^2}{\mu} \qquad \text{(by (45))}$$

$$\le \frac{\pi w_1(3 + w_1)r^2}{\mu}(m + 1)^2$$

$$\le 4\pi\frac{Dr}{\mu} < 4\pi\varepsilon' \qquad (49)$$

(by (47), (30), and (36)).

In addition, it follows from (37) and (45)-(47) that

$$\sum_{j=m+1}^{J_i} \Pr(D_{ij}) \cdot \Pr(D_{ij}^* | D_{ij})$$

$$\le \sum_{j=m+1}^{J_i} \frac{\pi(2j + 1)w_1(3 + w_1)r^2}{\mu^2} \frac{(3 + w_1)D^2}{(jw_1 - 3/2)}$$

$$\le \frac{16\pi}{\mu^2} \sum_{j=m+1}^{J_i} \frac{2j + 1}{j - (3/2w_1)} D^2r^2$$

$$\le \frac{16\pi}{\mu^2}(J_i - 1)D^2r^2 \frac{2(m + 1) + 1}{m + 1 - (3/2w_1)}$$

since $(J_i - 1)r \le D/w_1$:

$$\le \frac{16\pi}{\mu^2}\frac{D^3r}{w_1}\frac{2 + (1/(m + 1))}{1 - (3/2w_1(m + 1))}. \qquad (50)$$

We can assume that $m < J_i$. (If $m = J_i$, then (48) and (49) imply that $\Pr(F_i) \le 4\pi\varepsilon'$.) Thus, it follows from (47) that

$$\frac{3}{2w_1(m + 1)} = \frac{3}{2w_1\sqrt{D/w_1r}} = \frac{3}{2}\sqrt{\frac{r}{w_1D}}$$

$$= \frac{3}{2}\sqrt{\frac{Dr}{D^2w_1}} << 1 \qquad \text{(by (30))}.$$

Thus,

$$16\,\frac{2 + (1/(m + 1))}{1 - (3/2w_1(m + 1))} < 33,$$

so that a bound for (50) can be written as

$$\frac{33\pi}{\mu^2}\,\frac{D^3 r}{w_1} < \frac{33\pi D^2}{\mu}\,\varepsilon' \qquad \text{(by (30))}. \tag{51}$$

Combining (49) and (51), we obtain

$$\Pr(F_i) \le 37\pi\varepsilon' \max(D^2/\mu,\ 1) = \delta^4 \qquad \text{(see (31))}. \tag{52}$$

Now, let $c(S, h_u)$ be the expected capture time where the "resting" parameter $u$ of the hider's strategy $h_u$ is chosen to be $\delta\mu/2r$. If $F_n'$ is defined, as in (34), to be the event that capture does not occur before the hider leaves the point $Z_n$, then inequality (6) of Chapter 1 leads to the inequality

$$c(S, h_n) > u \sum_{n=1}^{\infty} \Pr(F_n') \ge u \sum_{n=1}^{N} \Pr(F_n') \tag{53}$$

(where $N$ is an integer which will be determined later)

$$\ge u \sum_{n=1}^{N} \left( \Pr\left( \bigcup_{i=1}^{n} E_i \right) - \sum_{i=1}^{n-1} \Pr(F_i) \right) \qquad \text{(see (35))}$$

$$\ge u \left( \sum_{n=1}^{N} \left(1 - \frac{2ru}{\mu}\right)^n - \sum_{n=1}^{N} (n - 1)\delta^4 \right)$$

(by (28) and (52))

$$\ge u \left( \frac{(1 - 2ru/\mu) - (1 - 2ru/\mu)^{N+1}}{2ru/\mu} - N^2\delta^4 \right).$$

Now if we choose $N = [1/\delta^2]$ and use $u = \delta\mu/2r$, we obtain

$$c(S, h_u) \ge \delta\frac{\mu}{2r}\left( \frac{1 - \delta - (1 - \delta)^{1/\delta^2}}{\delta} - 1 \right)$$

$$= \frac{\mu}{2r}\left(1 - 2\delta - (1 - \delta)^{1/\delta^2}\right). \tag{54}$$

It can be easily seen that for $0 < \delta < 1$ as assumed,

$$(1 - \delta)^{1/\delta^2} = \exp(1/\delta^2 \ln(1 - \delta)) < \exp(-1/\delta)$$
$$= \delta \exp(-\ln \delta - 1/\delta) < \delta \tag{55}$$

(since $\delta \cdot \ln \delta \geq -e^{-1} > -1$ so that $-\ln \delta - 1/\delta < 0$). Thus it follows from (54) and (55) that $c(S, h_u) \geq (\mu/2r)(1 - 3\delta)$. Q.E.D.

The following corollary will enable us to establish, in Section 4.5, the $\varepsilon$-optimality of $h_u$ for a more general cost function.

COROLLARY 2. If the hider uses the strategy $h_u$, then for any search trajectory and any positive $q$, the probability that the capture time $T$ exceeds $t$ satisfies for all $0 \leq t \leq q\mu/2r$

$$\Pr(T > t) \geq (1 - \varepsilon) \exp(-2rt/\mu)$$
$$\text{with} \quad \varepsilon \to 0 \quad \text{as} \quad r \to 0.$$

Proof. Let $I_t = [t/u] + 1$. Then (34), (35), (28), (52), and (32) imply that if the hider uses $h_u$, then the capture time $T$ satisfies for any search trajectory

$$\Pr(T > t) \geq \Pr(F'_{I_t}) \geq \Pr\left(\overline{\bigcup_{i=1}^{I_t} E_i}\right) - \sum_{i=1}^{I_t - 1} \Pr(F_i)$$
$$\geq (1 - 2ru/\mu)^{(t/u)+1} - \delta^4 t/u$$
$$= (1 - \delta)^{(2rt/\delta\mu)+1} - 2r\,\delta^3 t/\mu. \tag{56}$$

Let $\alpha$ be a positive number which satisfies

$$e^{-(1+\alpha)\delta} = 1 - \delta \qquad (\alpha \to 0 \quad \text{as} \quad \delta \to 0), \tag{57}$$

and let $\varepsilon$ satisfy

$$\varepsilon = \max(2(1 - (1 - \delta)e^{-\alpha q}), 2\delta^3 q e^q). \tag{58}$$

Then (56)-(58) imply that if $t = x\mu/2r$, where $0 \leq x \leq q$, then

$$\Pr(T > t) \geq (1 - \delta)e^{-(1+\alpha)x} - x\delta^3 \geq (1 - \delta)e^{-\alpha q}e^{-x} - q\delta^3$$

$$\geq (1 - \varepsilon/2)e^{-x} - \varepsilon e^{-q}/2 \geq (1 - \varepsilon)e^{-x}.$$

As $r \to 0$, both $\alpha$ and $\delta$ tend to zero, so that (58) implies that $\varepsilon \to 0$. Q.E.D.

It follows from Corollaries 1 and 2 that both the searcher and the hider can keep the probability of capture before time $t$ close to $1 - e^{-2rt/\mu}$. This is the expression obtained, in a different context, by Koopman (1956) for the probability of detection in a "random search."

## 4.4 EXTENSIONS OF THE PREVIOUS RESULTS

### Multidimensional Extensions

The results stated and proved in Sections 4.2 and 4.3 can be extended to any number of dimensions greater than 2, by the same techniques used in Theorems 1 and 2.

For example, if $Q$ is a three-dimensional compact region, then the construction of a search strategy which keeps the expected capture time below $(1 + \varepsilon)(\mu/\pi r^2)$ can be made as follows. Cover $Q$ by a large number $M$ of boxes of dimension $\alpha \times \alpha \times a$, where $r \ll \alpha \ll a$, randomly choose one of the boxes, and move along $N$ random horizontal segments which join the two faces of size $\alpha \times \alpha$, etc.

A hiding strategy which can keep the expected capture time above $(1 - \varepsilon)(\mu/\pi r^2)$ is the strategy $h_u$ presented by Definition 1 with a parameter $u$ which satisfies $1 \ll u \ll \mu/\pi r^2$.

### The Case of Several Searchers

We now consider a game in which J cooperative searchers seek a single mobile hider in a multidimensional region. We do not assume that all the searchers have the same characteristics so that each searcher i, i = 1,..., J, may have a different maximal speed $\nu_i$ and a different discovery radius $r_i$. Let $g_i$ be the maximal discovery rate of the i-th searcher ($g_i = 2r_i\nu_i$ for two dimensions, $\pi r_i^2 \nu_i$ for three dimensions, etc.). Then the total discovery rate $\bar{g}$ is defined as

$$\bar{g} = \sum_{i=1}^{J} g_i.$$

Under the assumption that $\max_{1 \le i \le J} r_i$ is small with respect to the dimension of Q and that the maximal velocity of the hider is not too small, it is possible to establish a result which is analogous to the one obtained for a single searcher; i.e., that $v \sim \mu/\bar{g}$. This follows from the fact that by adopting the strategy $h_u$, used in Section 4.3, the hider can keep the probability of capture after time t be close to the function $e^{-\bar{g}t/\mu}$. On the other hand, if each of the searchers adopts, independently, the strategy $s^*$ used in Section 4.2, then the probability of capture after time t will also be close to $e^{-\bar{g}t/\mu}$.

### The Case of an Arbitrary Starting Point

In Section 3.3, which dealt with a search for a mobile hider on a circle, we noted that the result $v = \mu/g$ depends on the assumption that the searcher has to start from a fixed point known to the hider and that if the starting point of the searcher is arbitrary, or if it is random, then the value would be substantially less than $\mu/g$. Obviously, that phenomenon does not occur in the multidimensional search, because both Theorems 1 and 2 are valid under any assumption about the location of the starting point, so that the result $v \sim \mu/g$ is not sensitive to the assumption $S(0) = O$.

## Search in Nonconvex Regions

The main result of Chapter 4, $v \sim \mu/g$, has been shown to hold under the assumption that the search space $Q$ is convex, but it seems to us that both the searcher's strategy $s^*$ presented in Section 4.2 and the hider's strategy $h_u$ presented in Section 4.3 can be modified for nonconvex regions in such a way that the basic results still hold for that case.

Considering the searcher's strategy, it is necessary to perturb $s^*$ near the boundary of $Q$, so that all the trajectories will pass inside $Q$. With regard to the hider's strategy, it seems reasonable to choose the points $Z_1, Z_2, Z_3, \ldots$ by the same method used by $h_u$, and the only problem is to make the movement from $Z_i$ to $Z_{i+1}$ in such a way that the probability of capture during this movement would still be negligible with respect to the probability of capture at $Z_i$. This requirement should be achievable if the detection radius is small enough.

## Nonhomogeneous Search Spaces

In Section 2.6, we have shown that the results established for the search game with an immobile hider can be extended to the case in which the maximal velocity of the searcher $\nu(Z)$ depends on the location of the searcher, and the discovery radius $r(Z)$ depends on the location of the hider. Using the framework and definitions of Section 2.6 (i.e., that $\nu(Z)$ is continuous and satisfies $0 < \nu_1 \leq \nu(Z) \leq \nu_2$ and $r(Z) = r_1\rho(Z)$, where $\rho(Z)$ is continuous and positive, and $r_1$ is small), we now extend the results of Sections 4.2 and 4.3 to this case.

In order to simplify the presentation, we describe this extension for two-dimensional regions. In this case, if $r_1$ is small, then the value of the game satisfies

$$v \sim \tau = \int_Q \frac{d\mu(Z)}{g(Z)} = \int_Q \frac{d\mu(Z)}{2\nu(Z)r(Z)},$$

where $\mu$ is the Lebesgue measure. The intuitive reason why the value should approximate $\tau$ is the following. Both the searcher and the hider can gurantee that, except for negligible time periods, the probability of detection in a small time interval $\Delta t$ is $(1 \pm \varepsilon)(\Delta t/\tau)$. Since both the searcher and the hider can keep the probability of detection close to an exponential function, it follows that the expected capture time is approximately $\tau$.

In order to achieve the value $\tau$, the hider's strategy $h_u$ should be modified as follows. The hiding points $Z_1, Z_2, \ldots$ should be chosen using the probability density $1/(\tau \cdot 2\nu(Z)r(Z))$, instead of the uniform density which was originally used.

With regard to the searcher, we first consider the simpler case in which $\nu(Z)$ is a constant, and only $r(Z)$ depends on $Z$. In this case, the only modification needed in $s^*$ is to construct the rectangles $Q_m$, $m = 1, \ldots, M$, in such a way that the width $\alpha_m$ of $Q_m$ would be proportional to $\bar{r}_m$, where $\bar{r}_m$ is the average of $r(Z)$ in $Q_m$. It is also required that $1 \gg a \gg \alpha_m \gg r_1$, where $a$ is the length of $Q_m$. If $\nu(Z)$ is not a constant, then a substantial modification is needed in $s^*$. In this case, the region $B_l$ in Figure 3 would not be a rectangle. Instead, it would be a strip as shown in Figure 8. The distance between any point $Z_1$ on the left boundary line and the corresponding

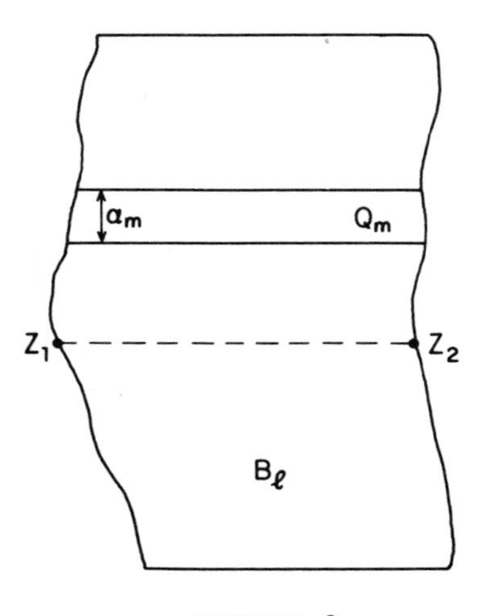


FIGURE 8

point $Z_2$ on the right boundary line (at the same level) should be equal to the maximal distance which can be traveled in a time interval of length a by a searcher moving from $Z_1$ in a horizontal direction. The region $B_l$ should then be divided into strips by parallel horizontal lines. The width $\alpha_m$ of the strip $Q_m$ should be proportional to $\bar{r}_m$, where $\bar{r}_m$ is the average of $r(Z)$ in $Q_m$. It should also be required that $l \gg a \gg \alpha_m \gg r_1$.

The fact that the above-described modifications of $h_u$ and $s^*$ guarantee the value $(1 \pm \varepsilon)\tau$ can be established by a technique similar to that used in Sections 4.2 and 4.3.

## 4.5 A GENERAL COST FUNCTION

It follows from Corollaries 1 and 2 of Sections 4.2 and 4.3 that each one of the players of the search game in a two-dimensional convex region can keep the probability of capture ater time t close to the function $e^{-\lambda t}$, where

$$\lambda = 2r/\mu. \tag{59}$$

We now wish to use those results in order to obtain the solution of the game in the case that the cost function U is a monotonic nondecreasing function of the capture time T, rather than T itself. In this case, one would expect that the value v of the game would satisfy

$$v \sim \lambda \int_0^\infty U(T) e^{-\lambda T}\, dT. \qquad (60)$$

We will show that (60) really holds, under the following assumption about U.

ASSUMPTION 1. There exist constants $\alpha > 0$, $\gamma \geq 0$, and $\beta > 1$ such that for all $T \geq \gamma$

$$U(\beta T) \leq \alpha U(T). \qquad (61)$$

Assumption 1 is satisfied for a wide variety of functions including all the polynomials, the bounded functions, and actually any reasonable function which does not increase faster than all polynomials.

In order to use the corollaries, we shall need to approximate the Laplace transform $f(\lambda)$ of $U(T)$,

$$f(\lambda) = \int_0^\infty U(T) e^{-\lambda T}\, dt, \qquad \lambda > 0, \qquad (62)$$

and the following lemma† shows that Assumption 1 is indeed the "right" condition needed for the desired approximation to hold.

LEMMA 2. If U(T) is a positive and nondecreasing function, then the following three conditions are equivalent:

†This lemma was suggested and proved by Abraham Ziv, IBM Israel Scientific Center.

(A) Assumption 1.

(B) $f(\lambda)$ is finite for all $\lambda > 0$, and for any $\lambda_0 > 0$ and $b > 1$, there exists a positive constant $\alpha$ such that

$$f(\lambda) \leq \alpha f(b\lambda) \qquad \text{for all} \quad 0 < \lambda \leq \lambda_0. \tag{63}$$

(C) $f(\lambda)$ is finite for all $\lambda > 0$, and for any $\lambda_0 > 0$ and $\varepsilon > 0$, there exists a positive number $q$ such that

$$\int_0^{q/\lambda} U(T)e^{-\lambda T}\, dT \geq (1 - \varepsilon) f(\lambda) \tag{64}$$

$$\text{for all} \quad 0 < \lambda \leq \lambda_0.$$

Proof. We shall show that (A) → (B) → (C) → (A).

First we show that (A) → (B).

If Assumption 1 holds and $U(T)$ is nondecreasing, then $U(T)$ satisfies for all $T \geq \gamma$,

$$U(T) \leq U(\gamma\beta^{[\ln(T/\gamma)/\ln\beta]+1})$$

$$\leq U(\gamma)\alpha^{(\ln(T/\gamma)/\ln\beta)+1} \leq \alpha_1 T^{\alpha_2},$$

where $\alpha_1$ and $\alpha_2$ are positive constants. Thus, the Laplace transform of $U$ is finite for all $\lambda > 0$.

Now

$$f(\lambda) = \int_0^\infty U(T)e^{-\lambda T}\, dT = \beta\int_0^\infty U(\beta x)e^{-\beta\lambda x}\, dx$$

$$\leq \beta\int_0^\gamma U(\beta x)e^{-\beta\lambda x}\, dx + \beta\int_\gamma^\infty \alpha U(x)e^{-\beta\lambda x}\, dx \qquad \text{(by (A))}$$

$$= \beta\int_0^\gamma (U(\beta x) - \alpha U(x))e^{-\beta\lambda x}\, dx + \beta\alpha f(\beta\lambda)$$

$$\leq \beta f(\beta\lambda)\left(\frac{1}{f(\beta\lambda)}\int_0^\gamma |U(\beta x) - \alpha U(x)|\, dx + \alpha\right).$$

Now, for any $0 < \lambda \leq \lambda_1$, $f(\beta\lambda) \geq f(\beta\lambda_1)$; thus

$$\sup_{0<\lambda\leq\lambda_1} \frac{f(\lambda)}{f(\beta\lambda)} \leq \beta\left(\frac{1}{f(\beta\lambda_1)} \int_0^{\gamma} |U(\beta x) - \alpha U(x)| \, dx + \alpha\right)$$

$$< \infty. \tag{65}$$

We have thus shown that (63) holds with $b = \beta$. We now show that it holds for all $b > 1$.

Let $n$ be a natural number such that $b < \beta^n$. Then for $\lambda_0 = \lambda_1/\beta^{n-1}$

$$\sup_{0<\lambda\leq\lambda_0} \frac{f(\lambda)}{f(b\lambda)} \leq \sup_{0<\lambda\leq\lambda_0} \frac{f(\lambda)}{f(\beta^n\lambda)} \leq \prod_{i=1}^{n} \sup_{0<\lambda\leq\lambda_0} \frac{f(\beta^{i-1}\lambda)}{f(\beta^i\lambda)}$$

$$\leq \left(\sup_{0<\lambda\leq\lambda_1} \frac{f(\lambda)}{f(\beta\lambda)}\right)^n < \infty \qquad \text{(by (65))},$$

which implies that (B) holds.

We now show that (B) $\rightarrow$ (C). Let

$$f_q(\lambda) = \int_0^{q/\lambda} U(T)e^{-\lambda T} \, dT. \tag{66}$$

Then, for any $b > 1$,

$$f(\lambda/b) \geq \int_{q/\lambda}^{\infty} U(T)e^{-\lambda T/b} \, dT \geq \int_{q/\lambda}^{\infty} e^{(b-1)q/b} U(T)e^{-\lambda T} \, dT$$

$$= (f(\lambda) - f_q(\lambda))e^{(b-1)q/b}.$$

Thus, it follows from (63) that for all $0 < \lambda \leq \lambda_0$

$$f(\lambda) - f_q(\lambda) \leq \alpha e^{-(b-1)q/b} f(\lambda),$$

which implies that for any $\varepsilon > 0$ one can choose $q$ to be large enough so that

$$f(\lambda) - f_q(\lambda) \leq \varepsilon f(\lambda) \qquad \text{for all } 0 < \lambda \leq \lambda_0.$$

Finally, we show that (C) → (A). Let q be a positive number. Then for any $0 < \lambda \leq \lambda_0$ and $\beta > 1$,

$$f(\lambda) \geq \int_{\beta q/\lambda}^{\infty} U(T) e^{-\lambda T}\, dT \geq U(\beta q/\lambda) e^{-\beta q}/\lambda.$$

On the other hand, it follows from (64) that

$$f(\lambda) \leq \frac{1}{1-\varepsilon} \int_{0}^{q/\lambda} U(T) e^{-\lambda T}\, dT \leq U(q/\lambda)/(1-\varepsilon)\lambda.$$

Thus, if $\beta > 1$, then for all $T > q/\lambda_0$

$$U(\beta T) \leq (e^{\beta q}/(1-\varepsilon))U(T),$$

so that Assumption 1 holds. Q.E.D.

We can now establish the $\varepsilon$-optimality of the strategies described in the previous sections for a general cost function.

THEOREM 3. If the cost function $U(T)$ is a positive non-decreasing function of the capture time $T$ which satisfies Assumption 1, then for any $\delta > 0$ there exists a detection radius $r_0$ so that for any search game with $r < r_0$ the strategies $s^*$ and $h_u$ presented in Sections 4.2 and 4.3 satisfy

$$v(s^*) \leq (1 + \delta)v \tag{67}$$

and

$$v(h_u) \geq (1 - \delta)v \tag{68}$$

with

$$v = \lambda \int_{0}^{\infty} U(T) e^{-\lambda T}\, dT \qquad \text{(see (59))}. \tag{69}$$

<u>Proof</u>. We establish (67) as follows. Corollary 1 implies that if the searcher uses $s^*$, then for any $\varepsilon > 0$, there exists a number $r_1$ so that for all $r < r_1$ and all $H$, the capture time is stochastically dominated by the measure which has density

$\lambda e^{-\lambda t/(1+\varepsilon)}$ $0 \leq t < \infty$. Thus

$$c(s^*, H) \leq \lambda \int_0^{\infty} U(T) e^{-\lambda T/(1+\varepsilon)} \, dT = \lambda f\left(\frac{\lambda}{1 + \varepsilon}\right), \tag{70}$$

where $f(\cdot)$ is the Laplace transform of U.

At the same time, it follows from Lemma 2 that for any $\lambda_1 > 0$ there exist constants $0 < a < 1$ and $\alpha > 1$ such that

$$f((1 - a)\lambda) \leq \alpha f(\lambda) \qquad \text{for all} \quad 0 < \lambda \leq \lambda_1. \tag{71}$$

Since $f(\lambda)$ is the Laplace transform of a positive function, then it is a convex function. Thus, for any $0 < \theta < 1$,

$$\begin{aligned} f((1 - \theta a)\lambda) - f(\lambda) &\leq \theta(f((1 - a)\lambda) - f(\lambda)) \\ &\leq \theta(\alpha - 1)f(\lambda) \qquad \text{(by (71))}. \end{aligned}$$

Thus

$$f((1 - \theta a)\lambda) \leq (1 + \theta(\alpha - 1))f(\lambda). \tag{72}$$

Now if we choose $\theta = \delta/(\alpha - 1)$ and $\varepsilon_0 = (1/(1 - \theta a)) - 1$, then there exists $r_2$ such that (70) holds with $\varepsilon = \varepsilon_0$ for all $r < r_2$, so that if we choose $r_0 = \min(r_1, r_2)$, then

$$c(s^*, h) \leq \lambda f\left(\frac{\lambda}{1 + \varepsilon_0}\right) \leq (1 + \delta)\lambda f(\lambda),$$

which establishes (67).

We now prove (68). Let $r_1$ be any positive number and let $\lambda_1 = 2r_1/\mu$. It follows from Lemma 2 that there exists a number $q_0$ so that

$$\int_0^{q_0/\lambda} U(T) e^{-\lambda T} \, dT \geq (1 - \delta/2) f(\lambda) \tag{73}$$

for all $0 < \lambda \leq \lambda_1$.

Now, choose $r_2$ to be small enough so that Corollary 2 is satisfied with $q = q_0$ and $\varepsilon = \delta/2$ for all $r \leq r_2$. Thus, it follows from Corollary 2 and (73) that if $r \leq r_0 = \min(r_1, r_2)$, then the expected cost is greater than or equal to

$$(1 - \delta/2)\lambda \int_0^{q_0/\lambda} U(T)e^{-\lambda T}\, dT \geq (1 - \delta)\lambda f(\lambda). \qquad \text{Q.E.D.}$$

## 4.6 SEARCH FOR AN INFILTRATOR

We conclude Chapter 4 by presenting some interesting unsolved problems which have some resemblance to the princess and monster game.

Assume that at time $t = 0$ an infiltrator enters the set Q through a known point O on the boundary of Q, and for all $t > 0$ moves inside Q (Figure 9). Suppose that the searcher has to defend a "sensitive zone" $B \subset Q$, so that he wishes to maximize the probability of capturing the infiltrator before he reaches the boundary of B (the infiltrator has the opposite goal). What are the optimal strategies of both players in this game? Such a problem with a relatively simple structure is the case in which Q is a narrow rectangle, as depicted in Figure 10.

A simpler, discrete problem with a similar flavor is the following. The search set is an array of n ordered cells. At time $t = 0$, both the searcher and the hider are located in cell number 1. At the end of each time unit, the searcher can move to any cell with distance not exceeding a certain integer

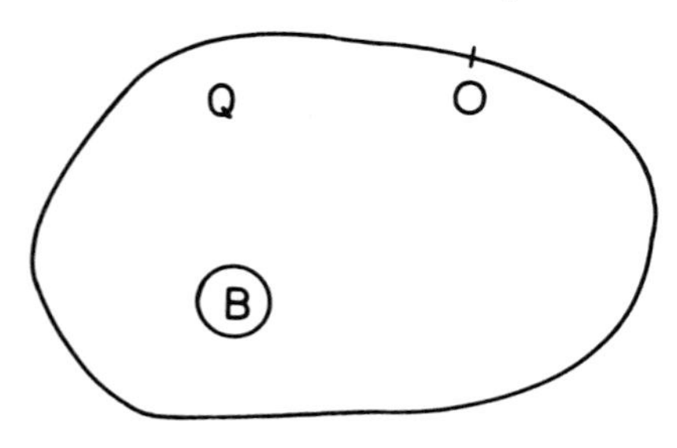


FIGURE 9

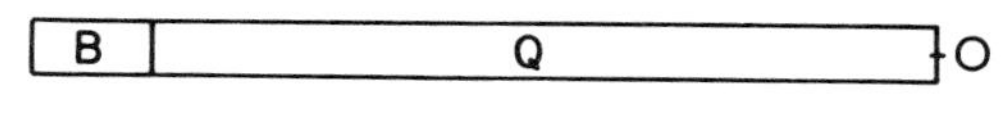


FIGURE 10

$k \geq 1$, while the hider can only move a distance of 1. Both players then stay at the chosen cell for the next time unit. The probability of capture is p for each time unit in which both players occupy the same cell (independently of previous history) and zero otherwise. The hider wins in the case that he reaches cell number n (in finite time) without getting captured, and loses otherwise. What are the optimal strategies of the players, and what is the asymptotic behavior of the probability that the hider wins, as p gets small and n gets large?

Another interesting problem of a similar type, which actually belongs to Part II, is the following.† The search space is the entire plane. A submarine (the hider) starts moving at $t = 0$ from the origin. The searcher starts moving at $t = 0$ from a fixed point different from the origin. The searcher wins if and only if he captures the submarine at some time $t > 0$. What are the optimal strategies of both players and what is the probability of ultimate capture? It is easy to verify that the probability P of ultimate capture satisfies

$$P \geq 1 - \prod_{i=0}^{\infty} \left(1 - \frac{r}{\pi R} \frac{1 - w}{w} \left(\frac{1 - w}{1 + w}\right)^{i}\right)$$

$$\geq 1 - \exp\left( - \left(\frac{r}{2\pi R} \frac{1 - w^2}{w^2}\right)\right), \tag{74}$$

†This problem was presented to me by Rufus Isaacs. Problems with a similar flavor were considered by Koopman (1946) and Danskin (1968).

where r is the capture radius, R is the distance of the searcher's starting point from the origin, and w, $w < 1$, is the maximal velocity of the hider. The searcher can achieve the probability of capture appearing in (74) by, at each of the time instants

$$t_i = R\left(\frac{1 + w}{1 - w}\right)^i, \qquad i = 0, 1, \ldots,$$

picking a random angle $\gamma_i$, $0 \leq \gamma_i < 2\pi$, where $\gamma_i$ has a uniform probability distribution and is independent of $\gamma_j$, $j < i$, moving in a straight line a distance $R_i = (w/(1 - w))t_i$ from the point O, and moving back to the point O.

It can easily be seen that in all the preceding problems, it is not a good policy for the hider to move in a straight line using his maximal velocity. A policy which does seem to be good for the hider is to move randomly for a certain period of time and only then to use his maximal velocity.

Remark 1. Search/evasion games involving a submarine (the evader) which tries to penetrate through a line guarded by the searcher, were considered by Agin (1967), Arnold (1962), Beltrami (1963), Houdebine (1963), Langford (1973), Lindsey (1968), and Pavillon (1963).

# PART II
# SEARCH GAMES IN UNBOUNDED DOMAINS

Chapter 5

# General Framework

## 5.1 ONE-DIMENSIONAL SEARCH GAMES

The work on search problems in unbounded domains was initiated by Bellman (1963) who introduced the following problem. An immobile hider is located on the real line according to a known probability distribution. A searcher, whose maximal velocity is one, starts from the origin O and wishes to discover the hider in minimal expected time. It is assumed that the searcher can change the direction of his motion without any loss of time. It is also assumed that the searcher cannot see the hider until he actually reaches the point at which the hider is located and the time elapsed until this moment is the duration of the game. This problem has been considered by Beck (1964, 1965) and Franck (1965). They found some properties of the optimal search trajectory, but a complete solution to the problem as presented by Bellman (1963) has not yet been found. Some other variations of this problem were considered by Beck and Warren (1973) and by Fristedt and Heath (1974) who used a different cost function, and by McCabe (1974) who considered the search for a random walker.

Beck and Newman (1970) presented a solution for the search on the real line considered as a game. As we already pointed out in the Introduction, if the capture time is used as the cost function and no restrictions are imposed on the hider, then the value of the game is infinite. Thus, in order to have a reasonable problem, one should either assume that the hider is in some way restricted, or that a normalized cost function is used. Beck and Newman used the first method and allowed the (immobile) hider to choose his location by using a probability distribution function h which has to belong to the class $Th_\lambda$ defined as follows.

DEFINITION 1. The class $Th_\lambda$ is the set of all hiding strategies in Q which satisfy the condition that the expected distance of the hiding point H from the origin is less than or equal to the constant $\lambda$. In other words, any $h \in Th_\lambda$ satisfies

$$\int_Q |H| \, dh \leq \lambda, \qquad (1)$$

where $|H|$ is the distance of H from the origin.

Such a restriction is the "natural" one for one-dimensional problems, because $C(S, H) \geq |H|$ for all S so that condition (1) is necessary in order to get a finite value for the game and, as we shall see in Chapter 7, condition (1) is also sufficient. We shall also see in Chapter 7 that the optimal search strategy does not depend on the constant $\lambda$, so that there is no need for the searcher to know the value of $\lambda$.

Another method of handling search games in unbounded domains, which is used by Gal (1974a) and Gal and Chazan (1976), is to impose no restriction on the hiding strategies but to normalize the cost function. Such a "natural" normalized cost function is the function $\widetilde{C}$ defined as

$$\widetilde{C}(S, H) = C(S, H)/|H|, \tag{2}$$

where C is the capture time. Since $C(S, H) \geq |H|$, it readily follows that the only normalization of the type $C/|H|^{\gamma}$ which yields a finite value is the normalization with $\gamma = 1$. This approach will be referred to as the "normalizing approach."

Actually, the two approaches just discussed lead to equivalent results under the following assumption which holds for all the games to be considered in Part II.

ASSUMPTION 1. For any positive constant $\beta$, there exists a mapping $\varphi_\beta$ of the search space Q into itself such that $\varphi_\beta(O) = O$, and for all $Z_1, Z_2 \in Q$

$$d(\varphi_\beta(Z_1), \varphi_\beta(Z_2)) = \beta\, d(Z_1, Z_2).$$

Under Assumption 1, any game with the restriction (1), with $\lambda$ being any positive constant, can be easily transformed into the game having the restriction (1) with $\lambda = 1$ by a construction which is similar to the one used in the scaling lemma of Chapter 1, with $\varphi_{1/\lambda}$.

We now show that, under Assumption 1, the game with the restriction (1) with $\lambda = 1$ is equivalent to the game with the normalized cost function (2). Let $h^*(H)$ be an $\varepsilon$-optimal hiding strategy which satisfies (1) with $\lambda = 1$. Then

$$\int_Q C(S, H)\, dh^*(H) \geq (1 - \varepsilon)v \quad \text{for all } S, \tag{3}$$

where v is the value of the game with restriction (1).

Since any hiding strategy with $E|H| < 1$ is dominated by a hiding strategy with $E|H| = 1$, we may assume that under $h^*$, $E|H| = 1$. Now, if we define the hiding strategy $\widetilde{h}(H)$ by

$$d\tilde{h}(H) = |H|\ dh^*(H),$$

then the strategy $\tilde{h}$ satisfies for all S

$$\int_Q \tilde{C}(S, H)\ d\tilde{h}(H) = \int_Q C(S, H)\ dh^*(H) \geq (1 - \varepsilon)v.$$

On the other hand, assume that $\tilde{h}$ is $\varepsilon$-optimal for the normalization approach; i.e.,

$$\int_Q \tilde{C}(S, H)\ d\tilde{h}(H) \geq (1 - \varepsilon)\tilde{v} \qquad \text{for all } S, \tag{4}$$

where $\tilde{v}$ is the value of the search game with the normalization approach. For all the one-dimensional problems considered in Part II, there exists a line which covers any $\varepsilon$-neighborhood of O, with length less than $\alpha\varepsilon$, where $\alpha$ is a bounded constant. Thus, there exist $\varepsilon$-optimal hiding strategies with $\Pr(|H| < \delta) = 0$ for some $\delta > 0$, and we assume that $\tilde{h}$ is such a strategy. Let

$$\int_Q \frac{d\tilde{h}(H)}{|H|} = a,$$

and define a hiding strategy $h_1(H)$ by

$$dh_1(H) = \frac{1}{a|H|}\ d\tilde{h}(H).$$

Then $h_1$ satisfies

$$\int_Q |H|\ dh_1(H) = \frac{1}{a}, \tag{5}$$

and for all S,

$$\int_Q C(S, H)\ dh_1(H) = \frac{1}{a}\int_Q \frac{C(S, H)}{|H|}\ d\tilde{h}(H)$$

$$\geq (1 - \varepsilon)\tilde{v}/a \qquad (\text{by } (4)).$$

Under Assumption 1, we can use the construction of the scaling lemma with the mapping $\varphi_a$ and obtain a hiding strategy

$h^*$, with $\int_Q |H| \, dh^*(H) = 1$, which makes sure that the expected capture time exceeds $(1 - \varepsilon)\tilde{v}$.

We have thus shown that $v = \tilde{v}$, so that the two approaches lead to equivalent results.

Remark 1. If Assumption 1 does not hold, then the two approaches may lead to different results. Such an example, in which Q is a half line, is discussed in Section 7.4.

The normalizing approach is more convenient for us to work with than the "restrictive approach," and thus we shall use the cost function $\tilde{C}$ given by (2). Working with the normalizing approach also has the following advantage: If $s^*$ is an $\varepsilon$-optimal strategy for the normalizing approach, then for any hiding strategy which satisfies $\int_Q |H| \, dh \leq 1$,

$$c(s^*, h) = \int_Q c(s^*, H) \, dh = \int_Q \frac{c(s^*, H)}{|H|} |H| \, dh$$

$$\leq \sup_H \frac{c(s^*, H)}{|H|} \leq (1 + \varepsilon)v.$$

Thus, $s^*$ is also $\varepsilon$-optimal for the restrictive approach. On the other hand, it can be easily seen that an $\varepsilon$-optimal search strategy for the restrictive approach need not be an $\varepsilon$-optimal strategy for the normalizing approach.

It should be noted that if the hider is mobile, then the appropriate restriction is $\int_Q |H(0)| \, dh \leq 1$, and the appropriate normalization is $\tilde{C} = C/H(0)$, where $H(0)$ is the location of the hider at time $t = 0$.

We shall deal mainly with one-dimensional search games and present only a brief discussion on multidimensional problems in Section 5.2. In Chapter 6, we prove some general minimax theorems which are used to obtain minimax trajectories for the search games on the infinite line solved in Chapter 7, and for the search games on a finite set of rays and in the plane which are presented in Chapter 8. The results presented in Chapters 6-8 are mainly based on the papers by Beck and Newman (1970), Gal and Chazan (1976), and Gal (1972, 1974a).[†]

## 5.2 MULTIDIMENSIONAL SEARCH GAMES

In this section, we discuss the framework of search games in the case that the search space is the entire N-dimensional Euclidean space with $N \geq 2$. The rules of the game are similar to those used in Part I. The searcher has to start from the origin O and moves along a continuous trajectory, with maximal velocity not exceeding 1. The hider is immobile and stays at a point H. The game terminates when the hider is within a distance of r $(r > 0)$ from the searcher. We shall be concerned with the appropriate restriction, which has to be imposed on the hider's strategies, in order to obtain a finite value for the game. In Section 5.1 we noted that condition (1) is necessary and sufficient to obtain a finite value for one-dimensional search games. Such a condition is not appropriate for multidimensional search games. In order to find the adequate moment restriction, we derive a necessary and sufficient condition for the finiteness of the value of the search game in the entire plane. Such a condition is

$$\int_Q |H|^2 \, dh \leq \lambda. \tag{6}$$

[†]Some of the material is taken with permission from SIAM J. Appl. Math 27 (1974), pp. 13-25 and 30 (1976), pp. 324-347. 

First we show that (6) is indeed a necessary restriction. If we allow hiding strategies with $\int_Q |H|^2 \, dh = \infty$, then the hider can use the strategy $h^*$ with the following probability density in the plane $H = (|H|, \theta)$, $0 \leq \theta < 2\pi$:

$$f(H) = \begin{cases} 1/\pi|H|^3 & \text{for } H \text{ such that } |H| \geq 1, \\ 0 & \text{otherwise.} \end{cases} \tag{7}$$

Since any trajectory S covers in each time interval [0, t] an area which does not exceed 2rt, it follows that the probability that the capture time exceeds t satisfies

$$\Pr(C(S, H) > t) > \int_{\sqrt{1+2rt/\pi}}^{\infty} \frac{2}{x^3} \, dx = \frac{\pi}{\pi + 2rt}.$$

Thus, for any S,

$$c(S, h^*) = \int_Q C(S, H) \, dh^* = \int_0^{\infty} \Pr(C(S, H) > t) \, dt$$

$$> \int_0^{\infty} \frac{\pi}{\pi + 2rt} \, dt = \infty,$$

which implies that $v = \infty$. On the other hand, if the hider's strategy has to satisfy condition (6), then the searcher can guarantee a finite capture time using the trajectory $S^*$ defined as follows. Start by moving in a straight line to the circle with radius 2r around the origin and encircle it, then move to the circle with radius 4r around the origin and encircle it, then to the circle with radius 6r, etc. The trajectory $S^*$ satisfies, for all H,

$$C(S^*, H) \leq |H| + r + 2\pi r \cdot \sum_{i=1}^{[|H|/2r]+1} 2i$$

$$< \frac{\pi}{2r} |H|^2 + (3\pi + 1)|H| + (4\pi + 1)r.$$

Since $E|H|^2 \le \lambda$, and consequently $E|H| \le \sqrt{E|H|^2} \le \sqrt{\lambda}$, it follows that for any admissible h

$$c(S^*, h) < (\pi/2r)\lambda + (3\pi + 1)\sqrt{\lambda} + (4\pi + 1)r,$$

so that the expected capture time guaranteed by $S^*$ is finite.

From the foregoing discussion, it follows that restriction (6) is the moment condition which has to be used in the case of an unbounded two-dimensional search. By using a similar argument, it is easy to show that the appropriate moment condition for N-dimensional unbounded search is $E|H|^N \le \lambda$. In the case that we use the approach of normalizing the capture time, a similar argument leads to the conclusion that the appropriate normalizing function is $|H|^N$. Thus, the normalized cost function should be $\tilde{C}(S, H) = C(S, H)/|H|^N$ for N-dimensional search games.

Chapter 6

# On The Optimality Of The Exponential Functions For Some Minimax Problems

## 6.1 INTRODUCTION

In Chapters 7 and 8, it is shown that, for the games to be discussed, finding minimax search trajectories amounts to a special case of finding a minimax search strategy for the following game.

The searcher chooses a sequence of positive numbers $X = \{x_j, I < j < \infty\}$, where $I > -\infty$ or $I = -\infty$, and the hider chooses an integer $i$. Define $X^{+i}$ as

$$X^{+i} = \{x_{j+i}, I < j < \infty\}. \tag{1}$$

Let the loss of the searcher $C(X, i)$ be given as

$$C(X, i) = F(X^{+i}), \tag{2}$$

where $F$ is a functional (i.e., a transformation from the space of all sequences into the real numbers) which satisfies several requirements to be specified in Section 6.2.

Let $A_a$ denote the geometric sequence

$$A_a = \{a^j, I < j < \infty\}. \tag{3}$$

Using the results proven in Appendix 2, we shall show, in Section 6.3, that a minimax search strategy can be found in the family of the geometric sequences $\{A_a, a > 0\}$.

In Section 6.4, we shall consider continuous versions of the problem and show that in this case, the minimax solution is an exponential function. This result will be used in Section 8.2.

Finally, in Section 6.5, we establish the uniqueness of the solution in the case that the functional F has a specific form which is encountered in Chapters 7 and 8.

## 6.2 REQUIRED PROPERTIES OF F FOR THE DISCRETE CASE

In order to simplify the notation, we shall assume that the integer I (appearing in (1)) is equal to 0. (The case where $I = -\infty$ will be discussed separately.)

At the first stage, we shall assume that F depends only on a finite number of terms of X. Thus, in this case, there is a positive integer k such that

$$F(X) = F(x_0, x_1, \ldots, x_k), \tag{4}$$

i.e.,

$$F(X^{+i}) = F(x_i, x_{i+1}, \ldots, x_{i+k}).$$

We require that F satisfy the conditions

$$F(X) \text{ is continuous} \quad \text{for all} \quad x_0 > 0,\ x_1 > 0, \ldots, x_k > 0, \tag{5}$$

$$F(\alpha X) = F(X) \qquad \text{for all} \quad \alpha > 0, \tag{6}$$

$$F(X + Y) \leq \max(F(X), F(Y)). \tag{7}$$

(For functionals which satisfy the homogenity condition (6), condition (7) is equivalent to the unimodality condition

$$F(\alpha X + (1 - \alpha)Y) \leq \max(F(X), F(Y)) \qquad \text{for all} \quad 0 \leq \alpha \leq 1.)$$

In addition, we assume that F satisfies the two conditions (which actually hold for all "reasonable" functionals which satisfy (5)-(7))

$$F(A_\infty) \overset{\text{def}}{=} \lim_{a\to\infty} F\left(\frac{1}{a^k}, \frac{1}{a^{k-1}}, \ldots, \frac{1}{a}, 1\right)$$

$$= \lim_{\varepsilon_k, \varepsilon_{k-1}, \ldots, \varepsilon_1 \to 0} F(\varepsilon_k, \varepsilon_{k-1}, \ldots, \varepsilon_1, 1) \tag{8}$$

and

$$F(A_0) \overset{\text{def}}{=} \lim_{a\to 0} F(1, a, \ldots, a^k)$$

$$= \lim_{\varepsilon_1, \varepsilon_2, \ldots, \varepsilon_k \to 0} F(1, \varepsilon_1, \varepsilon_2, \ldots, \varepsilon_k). \tag{9}$$

We now present two examples of F which satisfy conditions (5)-(9).

Example 1. Let $\alpha_0, \alpha_1, \ldots, \alpha_k$ be any real numbers and $\beta_0, \beta_1, \ldots, \beta_k$ be nonnegative numbers such that $\alpha_0^2 + \beta_0^2 > 0$ and $\alpha_k^2 + \beta_k^2 > 0$. Define

$$F(X) = \sum_{j=0}^{k} \alpha_j x_j \Big/ \sum_{j=0}^{k} \beta_j x_j. \tag{10}$$

It can be easily verified that conditions (5)-(9) hold.

Example 2. Let $\alpha_0, \alpha_1, \ldots, \alpha_k$ be any real numbers such that $\alpha_0 \neq 0$ and $\alpha_k \neq 0$, let $\beta_1, \ldots, \beta_{k-1}$ be any nonnegative numbers, and let $a_0$, $a_1$, and $a_2$ be positive numbers. Define

$$F(X) = \frac{\sum_{j=0}^{k-2} \alpha_j \sqrt{a_0 x_j^2 + a_1 x_{j+1}^2 + a_2 x_{j+2}^2}}{\sum_{j=1}^{k-1} \beta_j x_j}.$$

It can be verified using Minkowski's inequality (see Hardy et al., 1952) that (7) holds. All the other conditions are obvious.

The following lemma is an immediate consequence of (6) and (7).

LEMMA 1. If F satisfies (6) and (7), X is a positive sequence and $\beta_0, \ldots, \beta_L$ is a set of nonnegative numbers. Define

$$Y = \sum_{i=0}^{L} \beta_i X^{+i} \qquad \text{(see (1))}. \tag{11}$$

Then

$$F(Y) \leqq \max_{0 \leqq i \leqq L} F(X^{+i}). \tag{12}$$

## 6.3 DISCRETE MINIMAX THEOREMS

We note that condition (6) implies that for any geometric sequence $A_a$ and any positive integer i,

$$F(A_a^{+i}) = F(a^i, a^{i+1}, a^{i+2}, \ldots)$$

$$= F(1, a, a^2, \ldots) = F(A_a). \tag{13}$$

We now state and prove the minimax theorem for the class of games described in Section 6.1, where F depends on a finite number of terms of X.

This result uses Theorem 1 of Appendix 2 which states that the cone spanned by the family $\{X^{+i}, 0 \leq i < \infty\}$ contains a sequence which is "close enough" to a geometric sequence.

THEOREM 1. If F satisfies conditions (5)-(9), then for any positive sequence $X = \{x_i, 0 \leqq i < \infty\}$,

$$\overline{\lim_{i\to\infty}} F(X^{+i}) \geqq \inf_{0<a<\infty} F(A_a). \tag{14}$$

Theorem 1 claims that an optimal (or an $\varepsilon$-optimal) strategy for the class of games defined in Section 6.1 can be found among the geometric sequences $A_a$. This follows from the fact that if Theorem 1 is true, then (13) implies that for any positive sequence X,

$$\sup_{0 \leqq i < \infty} F(X^{+i}) \geqq \overline{\lim_{i\to\infty}} F(X^{+i})$$

$$\geqq \inf_{0<a<\infty} F(A_a) = \inf_{0<a<\infty} \sup_i F(A_a^{+i}).$$

Thus

$$\inf_X \sup_i F(X^{+i}) = \inf_{0<a<\infty} \sup_i F(A_a^{+i}) = \inf_{0<a<\infty} F(A_a).$$

Proof of Theorem 1. Let us denote

$$q = \inf_{0<a<\infty} F(A_a). \tag{15}$$

If (14) does not hold, then there exists a positive sequence $X = \{x_i,\ 0 \leqq i < \infty\}$, an integer $i_0$, and a positive number $\varepsilon$ such that for all $i \geqq i_0$,

$$F(X^{+i}) \leqq q - \varepsilon. \tag{16}$$

Without loss of generality, we may assume that $i_0 = 0$, so that (16) holds for all $i \geqq 0$.

Let $a = \overline{\lim}_{n\to\infty} x_n^{1/n}$. If $0 < a < \infty$, then Theorem 1 of Appendix 2 implies that for any $\delta > 0$ there exist nonnegative numbers $\beta_0, \beta_1, \ldots, \beta_L$ such that the sequence

$$Y = \{y_j,\ 0 \leqq j < \infty\} = \sum_{i=0}^{L} \beta_i X^{+i} \tag{17}$$

satisfies

$$|y_j - a^j| < \delta \qquad \text{for} \quad 0 \leqq j \leqq k. \tag{18}$$

It follows from Lemma 1 and (16) that

$$F(y_0, y_1, \ldots, y_k) = F(Y) \leqq \sup_{0 \leqq i < \infty} F(X^{+i}) \leqq q - \varepsilon. \tag{19}$$

Using (5) and (18), we obtain

$$F(A_a) = F(1, a, \ldots, a^k) \leqq q - \varepsilon,$$

which contradicts (15). Thus Theorem 1 has been proved for the case where $0 < a < \infty$.

If $a = \infty$ and (16) holds, then Theorem 1 of Appendix 2 implies that for any $\delta > 0$, we can find a sequence Y, defined by (17), satisfying $y_k = 1$ and $y_j < \delta$ for $0 \leqq j \leqq k - 1$. This, together with (8) and (19), implies that

$$\underline{\lim}_{a \to \infty} F\left(\frac{1}{a^k}, \ldots, \frac{1}{a}, 1\right) \leqq q - \varepsilon,$$

which contradicts (15).

A similar argument can be used for the case where $a = 0$. This completes the proof of Theorem 1.

We now illustrate the use of Theorem 1 by a simple example. Let F be defined as

$$F(X) = \frac{x_0 + x_1 + x_4 + x_5}{x_0 + 2x_3 + x_5}.$$

It follows from Theorem 1 that for any positive sequence $X = \{x_i,\ 0 \leqq i < \infty\}$,

$$\begin{aligned} \sup_{0 \leqq i < \infty} F(X^{+i}) &= \sup_{0 \leqq i < \infty} \frac{x_i + x_{i+1} + x_{i+3} + x_{i+4}}{x_i + 2x_{i+2} + x_{i+4}} \\ &\geqq \inf_{0 < a < \infty} \frac{(1/a^2) + (1/a) + a + a^2}{(1/a^2) + 2 + a^2} \\ &= \inf_{0 < a < \infty} F(A_a). \end{aligned} \tag{20}$$

It can easily be verified that $F(A_1) = 1$, $F(A_\infty) = 1$, and $F(A_0) = 1$, and since $(1/a) + a \geqq 2$, $F(A_a) > 1$ for any positive $a$ not equal to 1. Thus the minimal loss is 1; an optimal strategy for the first player is the sequence $X = \{1, 1, 1,\ldots\}$. The ε-optimal strategies can be obtained by using $X = \{1, a, a^2,\ldots\}$, $a$ being very large, or very small, or close to 1.

We now consider the case when $F(X)$ may depend on an infinite number of terms of $X$.

THEOREM 2. Let $\{F_k(X)\}$ be a sequence of functionals $F_k(X) = F_k(x_0, x_1,\ldots, x_k)$ defined for all $k > k_0$ ($k_0$ being any positive integer). Assume that $F_k$ satisfies conditions (5)-(9), and that for any positive sequence $X = \{x_i, 0 \leqq i < \infty\}$

$$F_{k+1}(x_0, x_1,\ldots, x_k, x_{k+1}) \geqq F_k(x_0, x_1,\ldots, x_k). \tag{21}$$

For any positive sequence $X$, define

$$F(X) = \lim_{k\to\infty} F_k(X), \tag{22}$$

$$F(A_\infty) = \lim_{k\to\infty} F_k(A_\infty) \quad \text{and} \quad F(A_0) = \lim_{k\to\infty} F_k(A_0) \tag{23}$$

(see (8) and (9)). Then

$$\overline{\lim_{i\to\infty}}\, F(X^{+i}) \geqq \inf_{0\leqq a\leqq\infty} F(A_a). \tag{24}$$

It should be noted that the inf in the right-hand side of (24) contains the end points $a = 0$ and $a = \infty$ because it may happen that

$$F(A_\infty) < \inf_{0<a<\infty} F(A_a).$$

This result is discussed by Gal (1972, pp. 42-43).

Proof of Theorem 2. Let

$$a = \overline{\lim_{n\to\infty}} x_n^{1/n}.$$

It has been established by Theorem 1 that for any integer k,

$$\overline{\lim_{i\to\infty}} F_k(X^{+i}) \geqq F_k(A_a).$$

Thus, it follows from (21) and (22) that for all k,

$$\overline{\lim_{i\to\infty}} F(X^{+i}) \geqq F_k(A_a). \tag{25}$$

Since $F(A_a)$ is defined by (22) (or (23)), then it follows from (25) that

$$\overline{\lim_{i\to\infty}} F(X^{+i}) \geqq F(A_a) \geqq \inf_{0\leqq a\leqq\infty} F(A_a). \qquad \text{Q.E.D.}$$

We now illustrate the use of Theorem 2.

Let L be a positive integer, $\beta_j$, $0 \leqq j \leqq L$, a nonnegative sequence, and $\alpha_j$, $0 \leqq j < \infty$, a sequence of real numbers satisfying $\alpha_j \geqq 0$ for $j > k_0$, where $k_0$ is any positive integer. Define

$$F(X) = \sum_{j=0}^{\infty} \alpha_j x_j \bigg/ \sum_{j=0}^{L} \beta_j x_j. \tag{26}$$

Then

$$\overline{\lim_{i\to\infty}} F(X^{+i}) \geqq \inf_{0\leqq a\leqq\infty} \sum_{j=0}^{\infty} \alpha_j a^j \bigg/ \sum_{j=0}^{L} \beta_j a^j.$$

This result is established by using Theorem 2 with

$$F_k(X) = \sum_{j=0}^{k} \alpha_j x_j \bigg/ \sum_{j=0}^{L} \beta_j x_j \qquad (k > k_0).$$

Remark 1. Condition (21) in Theorem 2 can be omitted, and then instead of (24), we obtain

$$\overline{\lim_{k\to\infty}}\ \overline{\lim_{i\to\infty}}\ F_k(X^{+i}) \geqq \overline{\lim_{k\to\infty}}\ \inf_{0<a<\infty} F_k(A_a). \tag{27}$$

The proof of (27) is similar to the proof of Theorem 2.

Theorem 2 can be extended to the case where X is a positive sequence infinite on both sides, i.e.,

$X = \{x_i, -\infty < i < \infty\}$.

In this case, if $F_k(X) = F(x_{-k}, \ldots, x_0, \ldots, x_k)$ satisfies (5)-(9) and a condition similar to (21) holds, then

$$\left.\begin{aligned} &\overline{\lim_{i\to+\infty}}\ F(X^{+i}) \geqq \inf_{0\leqq a\leqq\infty} F(A_a) \\ \text{and} \\ &\overline{\lim_{i\to-\infty}}\ F(X^{+i}) \geqq \inf_{0\leqq a\leqq\infty} F(A_a). \end{aligned}\right\} \tag{28}$$

The proof of this result is almost identical to the proof of Theorem 2 and we do not present it here.

Inequality (28) can be used in order to establish the following result.

COROLLARY 1. Let L be a positive integer, $\beta_j$, $-L \leqq j \leqq L$, a nonnegative sequence, and $\alpha_j$, $-\infty < j < \infty$, a sequence of real numbers satisfying $\alpha_j \geqq 0$ for $|j| > k_0$, where $k_0$ is any positive interger. Then

$$\overline{\lim_{i\to\pm\infty}} \frac{\sum_{j=-\infty}^{\infty} \alpha_j x_{i+j}}{\sum_{j=-L}^{L} \beta_j x_j} \geqq \inf_{0\leqq a\leqq\infty} \frac{\sum_{j=-\infty}^{\infty} \alpha_j a^j}{\sum_{j=-L}^{L} \beta_j a^j}.$$

Remark 2. Instead of considering the set of all positive sequences, we may consider only a subset B of it. Let $a_x = \overline{\lim}_{n\to\infty} x_n^{1/n}$. Then the modification of Theorem 1 (or Theorem 2) states that for any $X \in B$

$$\overline{\lim_{i\to\infty}} F(X^{+i}) \geq \inf_{X\in B} F(A_{a_x}).$$

The proof of this extension of Theorem 1 is based on the following argument. If $0 < a_x < \infty$, then Theorem 1 of Appendix 2 implies that there exist nonnegative constants $\beta_0, \beta_1, \ldots,$ such that $Y = \sum_{i=0}^{\infty} \beta_i X^{+i}$ is "close" to the geometric sequence $1, a_x, a_x^2, \ldots$ .

On the other hand, Lemma 1 implies that

$$\sup_{0\leq i<\infty} F(X^{+i}) \geq F(Y) \approx F(1, a_x, a_x^2, \ldots)$$

$$\geq \inf_{a_x : X\in B} F(1, a_x, a_x^2, \ldots).$$

The cases $a_x = 0$ and $a_x = \infty$ can be handled as in the proof of Theorem 1.

As an example, consider the expression

$$q = \inf_{X} \sup_{i>I} \frac{x_i + 100x_{i+2}}{x_{i+1}},$$

where $X = \{x_i, I < i < \infty\}$ belongs to the set B given by the condition

$$0 \leq R_1 < x_i < R_2 \qquad \text{for all} \quad I < i < \infty.$$

If $R_1 > 0$, then $a_x = 1$ for all $X \in B$. Thus, for both cases $I = -\infty$ and $I > -\infty$, $q = 101$ and $x_i$ = constant is a minimax solution.

If $R_1 = 0$, then we have to distinguish between two cases. If $I = -\infty$, then we still have $\overline{\lim}_{n\to-\infty} x_n^{1/n} \geq \lim_{n\to-\infty} R_2^{1/n} = 1$ for all $X \in B$, so that

$$q \geq \inf_{a\geq 1} \frac{1 + 100a^2}{a} \geq 101,$$

and the foregoing result still holds. On the other hand, if

$I > -\infty$, then $\{a_X : X \in B\} = \{a, 0 \leq a \leq 1\}$. Thus, $q = \inf_{0<a<1} (a^{-1} + 100a) = 101/10$ and $x_i = \alpha(1/10)^i$, where $0 < \alpha < R_2 10^{I+1}$ is a minimax solution.

A similar result holds for the continuous minimax theorems to be presented in the next section.

## 6.4 MINIMAX THEOREMS FOR THE CONTINUOUS CASE

In order to solve continuous problems of the type presented in Section 6.1, we extend the preceding results to the continuous case. The admissible strategies will belong to the following class of functions (for simplicity of exposition, we assume that I defined in Section 6.1 is equal to $-\infty$; however, all the results remain valid for finite I).

The class of all functions $X(\theta)$, $-\infty < \theta < \infty$, such that $X(\theta) > 0$ for all $\theta$ and, in addition, $X(\theta)$ is piecewise continuous and bounded on any finite interval, will be denoted by E. (29)

We define $X^{+t}$ by

$$X^{+t}(\theta) = X(\theta + t).$$

We shall be interested in

$$\inf_{X \in E} \sup_{t} F(X^{+t}),$$

where $F(X)$ is a functional defined for $X \in E$ which satisfies requirements (31)-(33) (below). At the first step, we shall assume that F depends only on the values of $X(\theta)$ in a finite segment $[-R, R]$, where R is a positive number. Thus we assume that

$$\text{If} \quad X(\theta) = Y(\theta) \quad \text{for} \quad -R \leqq \theta \leqq R, \quad \text{then} \quad F(X) = F(Y). \tag{30}$$

The following requirements are imposed on F.

F(X) is continuous with respect to the metric

$$d(X, Y) = \sup_{-R \leq \theta \leq R} |X(\theta) - Y(\theta)|, \tag{31}$$

$$F(\alpha X) = F(X) \quad \text{for any positive } \alpha, \tag{32}$$

and

$$F(X + Y) \leqq \max[F(X), F(Y)]. \tag{33}$$

A simple example of such an F is presented as follows.

Let A and B be two finite positive measures on the interval $0 \leqq \theta \leqq R$ and let

$$F(X) = \int_0^R X(\theta) \, dA(\theta) \Big/ \int_0^R X(\theta) \, dB(\theta). \tag{34}$$

Then F satisfies (30)-(33). (An additional example will be presented in Section 8.2.)

We shall use the following lemma which is a continuous version of Lemma 1.

LEMMA 2. Assume that F satisfies (30)-(33) and that $X \in E$ (see (29)). Let $\beta(t)$, $0 \leqq t \leqq L$, be a nonnegative continuous function. Define

$$Y(\theta) = \int_0^L \beta(t) X(t + \theta) \, dt. \tag{35}$$

Then

$$F(Y) \leqq \sup_{0 \leqq t \leqq L} F(X^{+t}). \tag{36}$$

Proof. For any finite closed interval, X is a piecewise continuous and bounded function and $\beta(t)$ is continuous. Thus, for any $\varepsilon > 0$, it is possible to find a sequence $0 = t_0 < t_1 < \ldots < t_n = L$, so that if we define

$$\beta_i = \int_{t_{i-1}}^{t_i} \beta(t)\, dt \quad \text{and} \quad Y_\varepsilon(\theta) = \sum_{i=1}^{n} \beta_i X(\theta + t_i), \tag{37}$$

then $Y_\varepsilon$ satisfies

$$\sup_{-R \leq \theta \leq R} |Y_\varepsilon(\theta) - Y(\theta)| < \varepsilon \quad \text{(see (35))}. \tag{38}$$

On the other hand, it follows from (32) and (33) that

$$F(Y_\varepsilon) \leqq \max_{1 \leqq i \leqq n} F(X^{+t_i}) \leqq \sup_{0 \leqq t \leqq L} F(X^{+t}). \tag{39}$$

Since F satisfies (31), then (38) and (39) imply (36).

Q.E.D.

Before presenting a continuous version of Theorem 1, we have to give conditions that replace (8) and (9).

We shall frequently use the notation

$$b_X \stackrel{\text{def}}{=} \ln\left(\overline{\lim_{n\to\infty}}\left(\int_n^{n+1} X(\theta)\, d\theta\right)^{1/n}\right). \tag{40}$$

The following additional conditions have to be satisfied:

$$F(e^{\infty\theta}) \stackrel{\text{def}}{=} \underline{\lim_{b\to\infty}} F(e^{b\theta})$$

$$\leqq \overline{\lim_{t\to\infty}} F(X^{+t}) \qquad \text{for any X such that } b_X = \infty, \tag{41}$$

$$F(e^{-\infty\theta}) \stackrel{\text{def}}{=} \underline{\lim_{b\to-\infty}} F(e^{b\theta})$$

$$\leqq \overline{\lim_{t\to\infty}} F(X^{+t}) \qquad \text{for any X such that } b_X = -\infty. \tag{42}$$

We now present a minimax theorem for the continuous case. The following result (as well as Theorem 4) uses Theorem 1´ of Appendix 2, which states that the cone spanned by the family $\{X^{+t}, 0 \leq t < \infty\}$ contains a function which is "close enough" to an exponential function.

THEOREM 3. Assume that F satisfies conditions (30)-(33), (41), and (42). Then for any $X \in E$ (see (29)),

$$\overline{\lim_{t\to\infty}} F(X^{+t}) \geqq \inf_{-\infty<b<\infty} F(e^{b\theta}) \stackrel{\text{def}}{=} q. \tag{43}$$

Proof. Let $X \in E$ and let $b_X$ be defined by (40). If $b_X = \infty$ or $b_X = -\infty$, then (43) is implied by (41) or (42); therefore assume that $-\infty < b_X < \infty$. If (43) does not hold, then there exists an $\varepsilon > 0$ such that for any $t > t_0$ (without loss of generality we may assume that $t_0 = 0$)

$$F(X^{+t}) < q - \varepsilon. \tag{44}$$

On the other hand, it follows from Theorem 1´ that for any $\delta > 0$, there exists a real number L and a nonnegative continuous function $\beta(t)$ so that the function $Y(\theta)$ defined by

$$Y(\theta) = \int_0^L \beta(t)X(t + \theta)\, dt$$

satisfies

$$\sup_{-R\leq\theta\leq R} |Y(\theta) - e^{b_X\theta}| < \delta. \tag{45}$$

It follows from Lemma 2 that

$$F(Y) \leqq \sup_{0\leqq t\leqq L} F(X^{+t}) < q - \varepsilon. \tag{46}$$

Hence, it follows from (31), (45), and (46) that

$$F(e^{b_x \theta}) \leqq q - \varepsilon,$$

which contradicts the definition of q. Q.E.D.

The following result which is used in Section 8.2 follows from Theorem 3.

COROLLARY 2. Let $F(X)$ be defined by (34), where A and B are two finite positive measures supported by the segments $[A_1, A_2]$ and $[B_1, B_2]$ such that the following condition is satisfied:

$$A_1 < B_1 \leqq B_2 < A_2. \tag{47}$$

Then

$$\overline{\lim_{t\to\infty}} F(X^{+t}) \geqq \inf_{-\infty<b<\infty} \int_0^R e^{b\theta}\, dA(\theta) \Big/ \int_0^R e^{b\theta}\, dB(\theta).$$

<u>Proof</u>. It has been pointed out that F satisfies conditions (30)-(33). We shall establish (41) by showing that

$$\text{if} \quad b_x = \infty \quad (\text{see } (40)), \qquad \text{then} \quad \overline{\lim_{t\to\infty}} F(X^{+t}) = \infty. \tag{48}$$

In fact, if (48) were not true, then there would exist a positive constant $\alpha$ so that for all $t > t_0$ (we can assume that $t_0 = 0$),

$$\int_{A_1}^{A_2} X(t + \theta)\, dA(\theta) < \alpha \int_{B_1}^{B_2} X(t + \theta)\, dB(\theta).$$

Condition (47) implies that there exists an $\varepsilon > 0$ so that

$$\int_{A_2-\varepsilon}^{A_2} X(t + \theta)\, dA(\theta) < \alpha \int_{A_1}^{A_2-2\varepsilon} X(t + \theta)\, dB(\theta).$$

Thus, for all positive u,

$$I_1 = \int_0^{u+\varepsilon}\left(\int_{A_2-\varepsilon}^{A_2} X(t+\theta)\, dA(\theta)\right) dt$$

$$< \alpha \int_0^{u+\varepsilon}\left(\int_{A_1}^{A_2-2\varepsilon} X(t+\theta)\, dB(\theta)\right) dt = I_2.$$

Using Fubini's theorem, we apply the following argument. $I_1$ is the integral of the function $X(\gamma)$ with respect to the product of Lebesgue measure and $A(\theta)$, over the set

$$\{(\theta, \gamma) : A_2 - \varepsilon \leqq \theta \leqq A_2,\ \theta \leqq \gamma \leqq \theta + u + \varepsilon\},$$

which contains the set

$$\{(\theta, \gamma) : A_2 - \varepsilon \leqq \theta \leqq A_2,\ A_2 \leqq \gamma \leqq u + A_2\}.$$

Thus

$$I_1 \geqq \left(\int_{A_2}^{u+A_2} X(\gamma)\, d\gamma\right) \int_{A_2-\varepsilon}^{A_2} dA(\theta).$$

Similarly,

$$I_2 \leqq \alpha\left(\int_{A_1}^{u+A_2-\varepsilon} X(\gamma)\, d\gamma\right) \int_{A_1}^{A_2-2\varepsilon} dB(\theta).$$

Thus it follows that there exists a positive constant $\alpha_1$ so that for all $u > \varepsilon$,

$$\int_{u+A_2-\varepsilon}^{u+A_2} X(\gamma)\, d\gamma < \alpha_1 \int_{A_1}^{u+A_2-\varepsilon} X(\gamma)\, d\gamma.$$

By using this inequality, it can be proved that

$$\overline{\lim_{n\to\infty}} \left(\int_n^{n+1} X(t)\, dt\right)^{1/n} < \infty,$$

which contradicts the assumption that $b_X = \infty$.

Thus (41) holds; a similar argument can be used to establish (42).

We are also interested in a subclass of E defined as follows:

$$\left.\begin{array}{l}\text{The class of all functions } X(\theta),\ -\infty < \theta < \infty, \text{ such} \\ \text{that } X(\theta) > 0 \text{ for all } \theta \text{ and in addition its deri-} \\ \text{vative } X'(\theta) \text{ is piecewise continuous and bounded} \\ \text{on any finite interval will be denoted by } E'.\end{array}\right\} \quad (49)$$

For this class we replace (31) by the following weaker condition: $F(X)$ is continuous with respect to the metric

$$d(X, Y) = \sup_{-R \leq \theta \leq R} (|X(\theta) - Y(\theta)| + |X'(\theta) - Y'(\theta)|). \quad (50)$$

Lemma 2 will be replaced by the following.

LEMMA 3. Assume that F satisfies (30), (50), (32), and (33) for all $X \in E'$ (see (49)). Let $\beta(t)$ be a nonnegative function having a continuous derivative. Define

$$Y(\theta) = \int_0^L \beta(t) X(t + \theta)\, dt.$$

Then

$$F(Y) \leq \sup_{0 \leq t \leq L} F(X^{+t}).$$

Proof. Note that

$$Y'(\theta) = \int_0^L \beta(t) X'(t + \theta)\, dt.$$

Thus, it follows from the fact that $\beta(t)$ is continuous and $X'$ is piecewise continuous and bounded on $[-R, R + L]$ that, as in (37), a function $Y_\varepsilon(\theta)$ can be found which satisfies (38) and $\sup_{-R \leq \theta \leq R} |Y'_\varepsilon(\theta) - Y'(\theta)| < \varepsilon$.

Hence, the method used for proving Lemma 2 is also applicable here.

We now present another version of Theorem 3 which will be used later.

THEOREM 4. Assume that F satisfies conditions (30), (50), (32), (33), (41), and (42). Then for any $X \in E'$ (see (49)).

$$\overline{\lim_{t\to\infty}} F(X^{+t}) \geqq \inf_{-\infty<b<\infty} F(e^{b\theta}).$$

One can establish the preceding result using an argument similar to the one used in Theorem 3. In this case, however, we require that $Y(\theta)$ satisfies (45) and

$$\sup_{-R\leq\theta\leq R} |Y'(\theta) - b_x e^{b_x\theta}| < \delta$$

(such a $Y(\theta)$ exists from Theorem 1′ of Appendix 2).

Using Lemma 3 we obtain the desired result.

We now consider the case where $F(X)$ may depend on values of X outside a finite interval.

THEOREM 5. Let $X \in E$ (resp. $E'$) and $\{F_K(x), K > K_0\}$ is a sequence of functionals satisfying the conditions of Theorem 3 (resp. Theorem 4) for $R = K$.

In addition assume that for all $X \in E$ (resp. $E'$),

$$F_n(X) \geqq F_m(X) \quad \text{for} \quad n > m. \tag{51}$$

For any $X \in E$ (resp. $E'$), define

$$F(X) = \lim_{K\to\infty} F_K(X), \tag{52}$$

$$F(e^{\infty\theta}) = \lim_{K\to\infty} F_K(e^{\infty\theta}), \tag{53}$$

and

$$F(e^{-\infty\theta}) = \lim_{K\to\infty} F_K(e^{-\infty\theta}). \tag{54}$$

Then

$$\overline{\lim_{t\to\pm\infty}}\ F(X^{+t}) \geqq \inf_{-\infty\leqq b\leqq\infty} F(e^{b\theta}). \tag{55}$$

Proof. We establish (55) for $t \to +\infty$ (the case of $t \to -\infty$ can be proven similarly).

Let $X \in E$ (resp. $E'$) and let $b_X$ be defined by (40). If $b_X = \infty$, then (41) implies that for all K,

$$\overline{\lim_{t\to\infty}}\ F_K(X^{+t}) \geqq F_K(e^{\infty\theta}).$$

Thus, it follows from (52) that for all K,

$$\overline{\lim_{t\to\infty}}\ F(X^{+t}) \geqq F_K(e^{\infty\theta}),$$

so that (55) follows from (53).

The case when $b_X = -\infty$ is similar. Thus, we may assume that $-\infty < b_X < \infty$.

It follows from Theorem 3 (resp. Theorem 4) that for any K,

$$\overline{\lim_{t\to\infty}}\ F_K(X^{+t}) \geqq F_K(e^{b_X\theta}).$$

Using (51) and (52), we obtain

$$\overline{\lim_{t\to\infty}}\ F(X^{+t}) \geqq F_k(e^{b_X\theta}).$$

Hence

$$\overline{\lim_{t\to\infty}}\ F(X^{+t}) \geqq F(e^{b_X\theta}) \geqq \inf_{-\infty\leqq b\leqq\infty} F(e^{b\theta}). \qquad \text{Q.E.D.}$$

Theorem 5 can be used for

$$F(X) = \int_{-\infty}^{\infty} X(\theta)\ dA(\theta) \Bigg/ \int_{-R}^{R} X(\theta)\ dB(\theta),$$

where A and B are two positive measures which satisfy condition (47). (In this case: $F_k(X) = \int_{-k}^{k} X(\theta)\ dA(\theta)/\int_{-R}^{R} X(\theta)\ dB(\theta)$.)

In particular, if A and B are probability measures which satisfy, for all real b,

$$\int_{-\infty}^{\infty} e^{bt}\, dA(t) \geqq \int_{-R}^{R} e^{bt}\, dB(t),$$

then for any $X \in E$,

$$\overline{\lim_{t\to\infty}} \left( \int_{-\infty}^{\infty} X(\theta + t)\, dA(\theta) \Bigg/ \int_{-R}^{R} X(\theta + t)\, dB(\theta) \right) \geqq 1.$$

## 6.5 UNIQUENESS OF THE MINIMAX STRATEGY

In this section we consider the case that $F(X)$ has the form

$$F(X) = \frac{\int_{-\infty}^{\infty} X(\theta)\, dA(\theta)}{X(0)}, \tag{56}$$

where $A(\theta)$ is a positive measure and $X(\theta) > 0$ for all $-\infty < \theta < \infty$. This case will be encountered in Chapters 7 and 8.

If F is given by (56), then the results established in Section 6.4 imply that

$$\inf_{X} \sup_{-\infty<y<\infty} \frac{\int_{-\infty}^{\infty} X(y + \theta)\, dA(\theta)}{X(y)} = \inf_{-\infty\leq b\leq\infty} \int_{-\infty}^{\infty} e^{b\theta}\, dA(\theta). \tag{57}$$

Thus, if the minimum of the right-hand side of (57) is obtained at $b = \bar{b}$, then

$$X(\theta) = \alpha e^{\bar{b}\theta}, \tag{58}$$

where $\alpha$ is a constant, is a minimax solution for the left-hand side of (57). We now investigate the conditions under which (58) is the unique solution of that problem. Such conditions are supplied by the following theorem.

THEOREM 6. Let $A(\theta)$ be a positive measure which is not concentrated at $\theta = 0$, and let $X(y)$, $-\infty < y < \infty$, be a positive function which is integrable on any finite interval. Let $f(b)$

be the bilateral Laplace transform of $A(\theta)$ defined

$$f(b) = \int_{-\infty}^{\infty} e^{b\theta}\, dA(\theta). \tag{59}$$

Assume that $f(b)$ attains its minimum at a point $-\infty < \bar{b} < \infty$, and that $f'(\bar{b})$ satisfies

$$\int_{-\infty}^{\infty} \theta e^{\bar{b}\theta}\, dA(\theta) = 0. \tag{60}$$

If

$$\sup_{-\infty<y<\infty} \frac{\int_{-\infty}^{\infty} X(y+\theta)\, dA(\theta)}{X(y)} \leq f(\bar{b}), \tag{61}$$

then

(a) If $A(\theta)$ is not arithmetic,[†] $X(y) = \alpha e^{\bar{b}y}$, where $\alpha$ is any positive constant.

(b) If $A(\theta)$ is arithmetic with span $\lambda$, $X(y) = \alpha(y)e^{\bar{b}y}$, where $\alpha(y)$ is a positive periodic function having period $\lambda$.

It should be noted that since $f(b)$ is a smooth convex function, then condition (60) is satisfied in the case that $\bar{b}$ is an interior point of the interval of convergence of $f(b)$ (but condition (60) may sometimes hold even when $\bar{b}$ is an end point of the interval of convergence of $f(b)$, or when $f(b)$ converges at a single point).

In order to establish Theorem 6 we first prove the following lemma.

[†] We use arithmetic in the sense used by Feller (1971), namely: a distribution A is arithmetic if it is concentrated on a set of points of the form $0, \pm\lambda \; \pm 2\lambda, \ldots$ . The largest $\lambda$ with this property is called the span of A.

LEMMA 4. Let $P(\theta)$, $-\infty < \theta < \infty$, be the distribution function of a probability measure which is not concentrated at $\theta = 0$, satisfying

$$\int_{-\infty}^{\infty} \theta \, dP(\theta) = 0, \tag{62}$$

and let $W(y)$, $-\infty < y < \infty$, be a positive function which is integrable on each finite segment. If

$$R(y) = \int_{-\infty}^{\infty} W(y + \theta) \, dP(\theta) \tag{63}$$

is defined for each y and if

$$R(y) \leq W(y) \qquad \text{for all} \quad -\infty < y < \infty, \tag{64}$$

then:

(a) If $P(\theta)$ is not arithmetic, then $W(y)$ is a constant.

(b) If $P(\theta)$ is arithmetic with span $\lambda$, then $W(y)$ is periodic having period $\lambda$.

Proof. Define a sequence $u_i$, $i = 1, 2, \ldots$, of independent random variables each of them having the distribution P. Let us denote

$$e_n = \sum_{i=1}^{n} u_i \tag{65}$$

and

$$w_n = -W(e_n). \tag{66}$$

Condition (64) implies that $w_n$ is a negative submartingale.[†] Hence, there exists a random variable $w^*$ such that

[†]A sequence of random variables $y_1, y_2, \ldots$ is called a submartingale if $E|y_n| < \infty$, $n = 1, 2, \ldots$, and $E(y_{n+1}|y_n, \ldots, y_1) \geq y_n$ a.s., $n = 1, 2, \ldots$ .

$$w_n \to w^* \quad \text{with probability 1} \tag{67}$$

(see Breiman, 1968, Chapter 5.4). We distinguish between two cases.

(a) $P(\theta)$ is not arithmetic.

In this case, (62) implies that the random walk defined by (65 visits every interval infinitely often, with probability 1. This, together with (67), implies that if $W$ is a continuous function, then it has to be a constant.

If $W$ is not continuous, we define

$$W_\varepsilon(y) = \frac{1}{2\varepsilon}\int_{-\varepsilon}^{\varepsilon} W(y+\theta)\,d\theta. \tag{68}$$

Thus defined, $W_\varepsilon$ is a continuous function satisfying the conditions of the lemma and so must be a constant $\alpha_\varepsilon$. It is easily verified that $\alpha_\varepsilon$ has the same value $\alpha$ for each $\varepsilon$, so that for all real $y$

$$\lim_{\varepsilon\to 0} W_\varepsilon(y) = \alpha. \tag{69}$$

On the other hand,

$$\lim_{\varepsilon\to 0} W_\varepsilon(y) = W(y) \quad \text{a.s. (almost surely)}, \tag{70}$$

which implies that $W(y) = \alpha$ a.s. This proves Lemma 4(a).

(b) $P(\theta)$ is arithmetic with span $\lambda$.

In this case, (62) implies that the random walk defined by (65) visits every point $j\lambda$ (where $j$ is any integer) infinitely often, with probability 1. Hence, (67) implies that $W(j\lambda)$ has the same value for each integer $j$.

In the same manner, we define $w_n = -W(\beta + e_n)$, where $\beta$ is any real number, and deduce that $W(\beta + j\lambda)$ has the same value for every integer j. Q.E.D.

Proof of Theorem 6. Define a probability distribution function $P(\theta)$ and a positive function $W(y)$ by

$$dP(\theta) = (e^{\bar{b}\theta}/f(\bar{b}))\, dA(\theta) \qquad \text{and} \qquad W(y) = X(y)e^{-\bar{b}y}.$$

Since $W(y)$ and $P(\theta)$ satisfy the conditions of Lemma 4, it follows that $W(y) = \alpha(y)$, where $\alpha(y)$ is a.s. a constant in case (a), and a periodic function in case (b). Q.E.D.

We now present an example which uses Theorem 6. Assume that the equation

$$\int_{\beta y}^{\infty} \frac{X(\theta)}{\theta}\, d\theta = \gamma X(y) \tag{71}$$

with $0 < \beta < 1$ holds for all $0 < y < \infty$. Making the change of variable $u = \ln \theta$ and defining $W(u) = X(e^u)$, we obtain the equation

$$\int_{R+\ln\beta}^{\infty} W(u)\, du = \gamma W(R), \qquad -\infty < R < \infty. \tag{71´}$$

Since

$$f(b) = \int_{\ln\beta}^{\infty} e^{b\theta}\, d\theta = -\frac{\beta^b}{b}$$

attains its minimum at $\bar{b} = 1/\ln \beta$, and $f(\bar{b}) = -e \ln \beta$, it follows from Theorem 6 that if $\gamma < -e \ln \beta$, then equation (71´), and consequently (71), has no positive solution.

If $\gamma = -e \ln \beta$, then equation (71´) has the unique solution $W(u) = \alpha e^{u/\ln\beta}$. Thus

$$X(y) = W(\ln y) = \alpha y^{1/\ln\beta},$$

where $\alpha$ is any positive constant.

If $\gamma > -e \ln \beta$, then it is easily verified that equation (71) has two solutions of the type $\alpha y^R$ (as well as many other solutions).

The discrete version of Theorem 6 is now given.

THEOREM 7. Let $A_j$, $-\infty < j < \infty$, be a nonnegative sequence, and denote

$$\varphi(a) = \sum_{j=-\infty}^{\infty} A_j a^j.$$

Assume that $\varphi(a)$ attains its minimum at a point $0 < \bar{a} < \infty$, so that

$$\sum_{j=-\infty}^{\infty} j A_j \bar{a}^{j-1} = 0. \tag{72}$$

If $X = \{x_i\}_{i=-\infty}^{\infty}$ is a positive sequence which satisfies for all $-\infty < i < \infty$

$$\frac{\sum_{j=-\infty}^{\infty} A_j x_{i+j}}{x_i} \le \varphi(\bar{a}), \tag{73}$$

then:

(a) If the span of $\{A_j\}$ is 1, then $x_i = \alpha \bar{a}^i$, where $\alpha$ is a positive constant.

(b) If the span of $\{A_j\}$ is $\lambda > 1$ (i.e., $A_j = 0$ for every i which is not a multiple of $\lambda$), then $x_i = \alpha_i \bar{a}^i$, where $\alpha_i$ is a positive periodic sequence with period $\lambda$.

*Proof*. Define a discrete measure $A(\theta)$ such that the measure of the point $\theta = j$ is $A_j$, and the measure of any interval not containing integral points is zero. Define $X(y)$ to be the step function $X(y) = x_i$ for $i - 1 < y \le i$, $-\infty < i < \infty$. Thus

defined, $A(\theta)$ and $X(y)$ satisfy the conditions of Theorem 6 with $\bar{b} = \ln \bar{a}$. Since $A(\theta)$ is arithmetic, it follows that $X(y) = \alpha(y)\bar{a}^y$, where $\alpha(y)$ is a periodic function with the period length equal to $\lambda$, the span of $A(\theta)$. Thus, if $\lambda = 1$, then $X(i) = x_i = \alpha\bar{a}^i$, and if $\lambda$ is an integer greater than 1, then $x_i = \alpha_i \bar{a}^i$, where $\alpha_i$ is a periodic sequence with period $\lambda$.
Q.E.D.

Note that condition (72) automatically holds if $\bar{a}$ is an interior point of the interval of convergence of $\varphi(a)$.

Remark 3. If condition (72) (resp. (60) for the continuous case) is not satisfied, then it is proven by Gal (1972) that the minimax strategy is no longer unique up to a multiplicative constant. Such a situation may occur only if the minimum of the bilateral Laplace transform $\varphi(a)$ is attained at a point $\bar{a}$ which is an end point of the interval of convergence of $\varphi(a)$ (or if $\varphi(a)$ converges only at a single point). For example, if the sequence $A_j$ satisfies $A_j = -1/j^3$, $-\infty < j \leq -1$, $A_1 = \beta > 0$, and all the other $A_j$ are zero, then $\varphi(a)$ converges for $a \geq 1$ and satisfies

$$\varphi(a) = \beta a + \sum_{j=1}^{\infty} \frac{a^{-j}}{j^3},$$

$$\frac{d\varphi(a)}{da} = \beta - \sum_{j=1}^{\infty} \frac{a^{-j-1}}{j^2}.$$

Since $\varphi(a)$ is a convex function, it follows that if

$$\frac{d\varphi(a)}{da} \geq 0 \quad \text{at} \quad a = 1, \tag{74}$$

then $\bar{a} = 1$, and if

$$\frac{d\varphi(a)}{da} < 0 \quad \text{at} \quad a = 1, \tag{75}$$

then $1 < \bar{a} < \infty$.

Inequality (74) is equivalent to $\beta \geq \sum_{j=1}^{\infty} 1/j^2$.

Thus, if $\beta < \sum_{j=1}^{\infty} 1/j^2$, then $\bar{a} > 1$ and (72) automatically holds. If $\beta = \sum_{j=1}^{\infty} 1/j^2$, then $\bar{a} = 1$ is an end point of the interval of convergence of $\varphi(a)$, but (72) holds, so that Theorem 7 is still valid.

On the other hand, if $\beta > \sum_{j=1}^{\infty} 1/j^2$, then (72) does not hold, and in this case there exist other functions, beside the constant sequences, which satisfy

$$\beta x_{i+1} + \sum_{j=1}^{\infty} \frac{x_{i-j}}{j^3} \leq \left(\beta + \sum_{j=1}^{\infty} \frac{1}{j^3}\right) x_i$$

for all $-\infty < i < \infty$.

Theorem 7 will be extensively used in Chapters 7 and 8. We now present a very simple example of its use. Let $A_1 = A_{-1} = 1/2$ and $A_j = 0$ for all other $j$. Then $\bar{a} = 1$ and $f(\bar{a}) = 1$. Thus, Theorem 7 implies for any positive sequence $X = \{x_j\}_{j=-\infty}^{\infty}$ that if $(x_{j+1} + x_{j-1})/2 \leq x_j$ for all $-\infty < j < \infty$, then $x_j =$ constant. This is a well-known result. If a positive sequence $x_j$, $-\infty < j < \infty$, is concave, then $x_j =$ constant.

It should be noted that, usually, the problem

$$\inf_X \sup_{i > i_0} \frac{\sum_{j=-\infty}^{\infty} A_j x_{i+j}}{x_i},$$

where $x_i > 0$, $A_j \geq 0$, and $\underline{i_0 > -\infty}$, does not have a unique (geometric) solution. For example, any sequence $Y = \{y_i\}$ which is strictly concave for $i_0 \leq i < \infty$, satisfies

$$\frac{y_{i-1} + y_{i+1}}{2y_i} < 1 \qquad \text{for all} \quad i > i_0,$$

and thus is a nonconstant solution to the problem

$$\inf_X \sup_{i>i_0} \frac{x_{i-1} + x_{i+1}}{2x_i}.$$

Still, Corollary 1 of Section 6.3 implies that any such sequence satisfies

$$\overline{\lim_{i\to\infty}} \frac{y_{i-1} + y_{i+1}}{2y_i} = 1.$$

Some other examples and a more detailed discussion are given by Gal (1972).

Remark 4. If we remove the assumption that the sequence X is positive, then the equation

$$x_{i-1} + x_{i+1} = \beta x_i, \qquad -\infty < i < \infty, \tag{76}$$

has an infinite number of solutions for all real $\beta$, including $\beta < 2$. This is so because starting with any $x_0$ and $x_1$, if we recursively define

$$x_{i+1} = \beta x_i - x_{i-1} \qquad \text{for all} \quad i \geq 1$$

and

$$x_{i-1} = \beta x_i - x_{i+1} \qquad \text{for all} \quad i \leq 0,$$

then such a sequence would satisfy (76).

A similar statement holds for all the other examples of Section 6.5. Thus, the requirement that X be positive is indispensable for all the results of Chapter 6.

Chapter 7

# Search On The Infinite Line

## 7.1 INTRODUCTION

In this chapter, we consider the search game on the infinite line already described in Section 5.1. We shall usually assume that the hider is immobile and that capture occurs the first time that the searcher passes the point occupied by the hider, but in Sections 7.5 and 7.6 we shall consider some variations of the game in which these assumptions do not hold. As usual, we assume that the searcher can use any continuous trajectory S which satisfies: $S(0) = 0$, for all $t > 0$ $S(t)$ is a real number, and for all $t_2 > t_1 \geq 0$ $|S(t_2) - S(t_1)| \leq t_2 - t_1$. The hider can choose any real number H, and the cost function $\tilde{C}(S, H)$ is given by

$$\tilde{C}(S, H) = C(S, H)/|H|, \tag{1}$$

where $C(S, H)$ is the capture time. In the case that mixed strategies are used, we shall denote the expected value of $\tilde{C}$ by $\tilde{c}$.

At first we shall identify some properties which a search trajectory must have in order to be "efficient." Such a property is obviously

$$\sup_{t>0} S(t) = \infty \qquad \text{and} \qquad \inf_{t>0} S(t) = -\infty, \tag{2}$$

because otherwise there exist hiding points H which are never discovered by S, i.e., $v(S) = \infty$. Moreover, any mixed search strategy s with $v(s) < \infty$ has to satisfy condition (2) with probability 1. Thus, it can be assumed that all the relevant mixed strategies make their probabilistic choice among the trajectories which satisfy (2).

We have already mentioned in Section 2.1 that, when searching for an immobile hider, it can be assumed that the velocity along the search trajectory is 1. Using a dominance argument of a similar type, it can be shown that it is sufficient to consider search trajectories $S(t)$ with $S'(t) = 1$ or $S'(t) = -1$ for all $t > 0$ except a denumerable set of "break points" defined as follows.

DEFINITION 1. The trajectory $S(t)$ has a positive break point (PBP) at the time $t_0 > 0$ if there exists an $\varepsilon > 0$ such that

$$S'(t) = +1 \quad \text{for} \quad t_0 - \varepsilon < t < t_0$$

and

$$S'(t) = -1 \quad \text{for} \quad t_0 < t < t_0 + \varepsilon.$$

The trajectory $S(t)$ has a negative break point (NBP) at $t_0$ if

$$S'(t) = -1 \quad \text{for} \quad t_0 - \varepsilon < t < t_0$$

and

$$S'(t) = +1 \quad \text{for} \quad t_0 < t < t_0 + \varepsilon$$

for some $\varepsilon > 0$.

Obviously, between any two PBPs there exists at least one NBP and between any two NBPs there exists at least one PBP. We also have the following property.

Let S be any search trajectory with $v(S) < \infty$. Then for any $t_0 > 0$:

(a) The number of the break points before $t_0$ is infinite.

(b) The number of the break points after $t_0$ is infinite.

Property (b) readily follows from (2), while property (a) follows from the fact that if there is a first positive break point, then the hider can obtain any large payoff by choosing $H = -\varepsilon$ with $\varepsilon$ being very small, while if there is a first negative break point, then the hider can obtain any large payoff by choosing $H = +\varepsilon$. Thus, the set of break points is a doubly infinite sequence.

Remark 1. In the approach that we use, every admissible search trajectory has to start by making an infinite number of small "oscillations" near O. This phenomenon follows from the normalization we use to define the cost function $\tilde{C}$. In order to obtain a "practical" solution which does have a first break point, one could either assume that there is a positive discovery radius r, so that all the points H with $|H| < r$ are discovered at $t = 0$, or one could use the restrictive approach discussed in Chapter 5. From the optimal search strategy that we shall present, one can immediately obtain an $\varepsilon$-optimal strategy for the above-mentioned versions of the game by simply ignoring all the break points before a certain instant $t_0$ where $t_0$ is very small. Such a solution will use trajectories which have a first break point and do not have small "oscillations."

From our previous conclusions it follows that any search trajectory S can be represented by a doubly infinite sequence $\{x_i\}_{i=-\infty}^{\infty}$ with the convention that each $x_i$ with an even i represents the location of a positive break point, while if i is an odd integer, then $-x_i$ represents the location of a negative break point. For any even i, the searcher moves from the point $x_i$ to $-x_{i+1}$ and then to $x_{i+2}$, etc. We now show that any reasonable search trajectory has to satisfy

$$0 < \dots < x_{2j-2} < x_{2j} < x_{2j+2} < \dots \tag{3}$$

and

$$\dots < -x_{2j+3} < -x_{2j+1} < -x_{2j-1} < \dots < 0, \tag{4}$$

i.e., both the positive break points and the negative break points are monotonic, and all the $x_i$ are positive.

Property (3) follows from the following consideration. If S is a trajectory which does not satisfy (3), then we can find a trajectory $\overline{S}$ which dominates S and satisfies (3) using the following construction. Let J be a subset of the even integers defined by the property

$$i \in J \quad \text{iff} \quad i \text{ is even} \quad \text{and } x_i > \max(\sup_{2j<i} x_{2j}, 0).$$

From the foregoing discussion, it follows that any trajectory S with $v(S) < \infty$ has to satisfy

$$\inf_{i \in J} i = -\infty \quad \text{and} \quad \sup_{i \in J} i = \infty.$$

We can order the elements of J by their magnitude so that

$$J = \{i_k\}_{k=-\infty}^{\infty} \quad \text{with} \quad i_{k+1} > i_k.$$

Now define the trajectory $\overline{S}$ by $\overline{S} = \{\overline{x}_k\}_{k=-\infty}^{\infty}$ with

$$\overline{x}_{2k} = x_{i_k} \quad \text{and} \quad \overline{x}_{2k+1} = \max_{i_k < 2j+1 < i_{k+1}} x_{2j+1}.$$

It is obvious that $\overline{S}$ dominates S and satisfies (3). Using a similar construction, we can start from $\overline{S}$ and obtain a trajectory $\overline{S}_1$ which dominates $\overline{S}$ and satisfies (3) and (4).

We have just shown that it is sufficient to consider search trajectories which satisfy (3) and (4). Such trajectories will be referred to as "periodic and monotonic."

## 7.2 THE MINIMAX SEARCH TRAJECTORY

In this section, we present the minimax pure strategy of the searcher. In other words, we find a trajectory $S^*$ which minimizes $v(S)$, where

$$v(S) = \sup_H \tilde{C}(S, H) = \sup_H \frac{C(S, H)}{|H|}.$$

First we note that by property (2) of Section 7.1, any hiding point H has to satisfy

$$x_i < H \leq x_{i+2} \qquad \text{for some even } i,$$

or

$$-x_{i+2} \leq H < -x_i \qquad \text{for some odd } i.$$

Since

$$C(S, H) = 2 \sum_{j=-\infty}^{i+1} x_j + |H|,$$

where $x_i < |H| \leq x_{i+2}$, then

$$v(S) = \sup_H \left(1 + 2 \frac{\sum_{j=-\infty}^{i+1} x_j}{|H|}\right) = 1 + 2 \sup_{-\infty<i<\infty} \frac{\sum_{j=-\infty}^{i+1} x_j}{x_i}.$$

We now use the results obtained in Chapter 6. It follows from Corollary 1 of Section 6.3 and Theorem 7 of Section 6.5 that

$$\inf_S \sup_{-\infty<i<\infty} \frac{\sum_{j=-\infty}^{i+1} x_j}{x_i} = \min_{a>0} \frac{\sum_{j=-\infty}^{i+1} a^j}{a^i} = \min_{a>1} \frac{a^2}{a-1} = 4$$

with the unique optimal sequence being $x_i = \alpha 2^i$, where $\alpha$ is any positive constant. (Note that $\alpha = \alpha_1$ and $\alpha = \alpha_1 4^j$, where j is any integer, represent the same trajectory.) Thus, the minimal cost assured by using a pure strategy (fixed trajectory) is

$$\overline{VP} = \inf_S v(S) = 9.$$

It should be noted that the trajectory $S^* = \{\alpha 2^i\}_{i=-\infty}^{\infty}$ is still minimax for the case that the capture time is used as the cost function and the hider's strategies has to satisfy $E|H| \le \lambda$ because

$$\sup_{h:E|H|\le\lambda} c(S^*, h) = \sup_{h:E|h|\le\lambda} \int \tilde{C}(S^*, H)|H|\, dh$$

$$\le \lambda \sup_H \tilde{C}(S^*, H) = 9\lambda,$$

while, on the other hand, any hiding strategy which has atoms of probability masses $\lambda/|H|$ and $1 - \lambda/|H|$ at the points H, ($|H| \ge \lambda$), and 0, respectively, is an admissible strategy. Thus

$$\inf_S \sup_{h:E|H|\le\lambda} c(S, h) \ge \inf_S \sup_{H:|H|\ge\lambda} \frac{\lambda}{|H|} C(S, H)$$

$$= \lambda \inf_S \sup_{H:|H|\ge\lambda} \tilde{C}(S, H)$$

$$\ge \lambda\left(1 + 2 \sup_{i:x_i\ge\lambda} \frac{\sum_{j=-\infty}^{i+1} x_j}{x_i}\right)$$

since $x_i \to \infty$ as $i \to \infty$:

$$\ge \lambda\left(1 + 2 \overline{\lim_{i\to\infty}} \frac{\sum_{j=-\infty}^{i+1} x_j}{x_i}\right)$$

by Corollary 1 of Section 6.3:

$$\ge 9\lambda.$$

We shall reiterate this subject in Section 7.4.

## 7.3 OPTIMAL STRATEGIES

In the previous section, we showed that if the searcher wishes to use a fixed trajectory which assures him a minimal cost, then he should use the "geometric" trajectory $\{\alpha 2^i\}_{i=-\infty}^{\infty}$. We now assume that the searcher is satisfied with minimizing the expected cost, and we find an optimal mixed strategy $s^*$ which achieves this goal. We shall use the same method adopted by Beck and Newman (1970) to demonstrate that $s^*$ belongs to the family $A = \{s_a\}$ of mixed strategies defined as follows.

DEFINITION 2. For any $a > 1$, the strategy $s_a$ chooses a trajectory $S = \{x_i\}_{i=-\infty}^{\infty}$ with $x_i = a^{i+u}$, where u is uniformly distributed in the interval $[0, 2)$.

The strategy $s_a$ is a random choice among the geometric trajectories with rate of increase a. (The random variable u is restricted to $[0, 2)$ because S is a periodic function of u with period 2.) Each strategy $s_a$ has the following property.

LEMMA 1. For any hiding point H,

$$\tilde{c}(s_a, H) = 1 + \frac{a + 1}{\ln a}. \tag{5}$$

Proof. Assume for convenience that $H > 0$. Then for any trajectory $S = \{x_i\}_{i=-\infty}^{\infty}$, the capture time $C(S, H)$ satisfies

$$C(S, H) = H + 2 \sum_{j=-\infty}^{i+1} x_i,$$

where $x_i < H \leq x_{i+2}$ and i is even. Thus

$$c(s_a, H) = H + 2E\left( \sum_{j=-\infty}^{i+1} x_j \middle| x_i < H \leq x_{i+2} \right)$$

by Definition 2:

$$= H + 2E\left(\frac{a}{1 - (1/a)} x_i \middle| x_i < H \le a^2 x_i\right)$$

$$= H + \frac{2a^2}{a-1} E\left(x_i \middle| x_i < H \le a^2 x_i\right)$$

$$= H + \frac{2a^2}{a-1} \int_0^2 \frac{1}{2} Ha^{u-2}\, du = H\left(1 + \frac{a + 1}{\ln a}\right),$$

which establishes (5).

Let

$$q = \min_{a>1}\left(1 + \frac{a + 1}{\ln a}\right) = 1 + \frac{\bar{a} + 1}{\ln \bar{a}}. \tag{6}$$

Since $(a + 1)/\ln a$ is a unimodal function for $a > 1$, (If $0 \le \theta \le 1$, then

$$\frac{\theta a_1 + (1 - \theta)a_2 + 1}{\ln(\theta a_1 + (1 - \theta)a_2)} \le \frac{\theta(a_1 + 1) + (1 - \theta)(a_2 + 1)}{\theta \ln a_1 + (1 - \theta) \ln a_2}$$

$$\le \max\left(\frac{a_1 + 1}{\ln a_1}, \frac{a_2 + 1}{\ln a_2}\right).)$$

It follows that its derivative vanishes at $\bar{a}$. Thus

$$\frac{\bar{a} + 1}{\ln \bar{a}} = \bar{a} \tag{7}$$

and

$$q = 1 + \bar{a}. \tag{8}$$

By using $s_{\bar{a}}$, the searcher can make sure that the expected cost does not exceed $q$. A numerical calculation of $q$ shows that $q \approx 4.6$ which is a little more than half of the minimax cost guaranteed by the minimax search trajectory presented in Section 7.2. We now show that $s_{\bar{a}}$ is indeed an optimal strategy by proving that the value of the game is $q$. The proof is constructive and presents an $\varepsilon$-optimal strategy of the hider.

THEOREM 1. For any $\varepsilon > 0$, there exists a hiding strategy $h_\varepsilon$ such that for all S

$$\tilde{c}(S, h_\varepsilon) \geq (1 - \varepsilon)q,$$

where q is given by (6).

Proof. Denote

$$R = e^{(1-\varepsilon)/\varepsilon}, \tag{9}$$

and let $h_\varepsilon$ be the strategy which chooses the hiding point H using the probability distribution function G defined as

$$\left.\begin{aligned}
G(H) &= 0 && \text{for } H < -R \\
G(-R) &= \varepsilon/2, && \\
G(H) &= (1 + \ln R - \ln|H|)\varepsilon/2 && \text{for } -R \leq H \leq -1, \\
G(H) &= 1/2 && \text{for } -1 \leq H \leq 1, \\
G(H) &= 1/2 + (\ln H)\varepsilon/2 && \text{for } 1 \leq H < R, \\
G(H) &= 1 && \text{for } H \geq R
\end{aligned}\right\} \tag{10}$$

($h_\varepsilon$ has density $\varepsilon/2|H|$ for $1 \leq |H| < R$ and has two probability atoms, of mass $\varepsilon/2$ each, at $H = -R$ and $H = R$).

Since the hiding point H satisfies $1 \leq |H| \leq R$ and $|H| = R$ with a positive probability, then we can obviously consider (against $h_\varepsilon$) only search trajectories which have a first break point and a last break point. Let S be such a trajectory. Then we can assume that S starts by going from 0 to $x_0$ and back to 0, then to $-x_1$ and back to 0, etc. The last two steps of S are of length R each. Thus S can be represented by a set of positive numbers $\{x_i\}_{i=0}^{n}$ with $x_{i+2} > x_i$, $0 \leq i \leq n - 2$. We denote for convenience $x_{-1} = x_{-2} = 1$. The trajectory S satisfies

$$\tilde{c}(S, h_\varepsilon) = \sum_{i=0}^{n} \int_{x_{i-2}}^{x_i} \left(2 \sum_{j=0}^{i-1} x_j + H\right) \frac{dG(H)}{H}$$

$$= 1 + 2 \sum_{i=0}^{n-1} x_i \left( \int_{x_{i-1}}^{R} \frac{dG(H)}{H} + \int_{x_i}^{R} \frac{dG(H)}{H} \right)$$

by (10):

$$= 1 + \varepsilon \sum_{i=0}^{n-1} x_i \left( \frac{1}{x_i} + \frac{1}{x_{i-1}} \right)$$

$$\geq 1 + \varepsilon n \left( 1 + \left( \prod_{i=0}^{n-1} \frac{x_i}{x_{i-1}} \right)^{1/n} \right) = 1 + n\varepsilon (1 + R^{1/n})$$

by (9):

$$= 1 + (1 - \varepsilon) \frac{1 + R^{1/n}}{\ln(R^{1/n})} > (1 - \varepsilon) \left( 1 + \frac{1 + R^{1/n}}{\ln(R^{1/n})} \right)$$

by (6):

$$\geq (1 - \varepsilon) q. \qquad \text{Q.E.D.}$$

In Appendix 1 we show for search games in general that the searcher has an optimal strategy and that the game has a value (which implies that the hider has $\varepsilon$-optimal strategies). However, the hider need not have an optimal strategy. An example in which there is no optimal strategy for the hider is the search game with a mobile hider presented in Section 3.2. We now show that the same phenomenon occurs in the present game. The intuitive reason for this fact is that if an optimal hiding strategy would have existed, then it would have to use a probability density which is proportional to $1/|H|$ for all $0 < |H| < \infty$. However, such a probability density, obviously, does not exist.

THEOREM 2. Any hiding strategy h satisfies

$$\inf_s \tilde{c}(s, h) < q,$$

where q is given by (6).

Proof. Assume that there exists a hiding strategy $h^*$ which satisfies

$$\inf_s \tilde{c}(s, h^*) \geq q. \tag{11}$$

Let $H^*$ be the random variable which corresponds to $h^*$. It can be easily verified that the random variable $\beta H^*$, where $\beta$ is any constant, also corresponds to a strategy which satisfies (11). Since a probability mixture of hiding strategies which satisfy (11) also satisfies (11), it follows that if R is a random variable which has a density with respect to the Lebesgue measure, then the random variable $H_1$ defined by

$$H_1 = \begin{cases} RH^* & \text{with probability } 1/2, \text{ and} \\ -RH^* & \text{with probability } 1/2 \end{cases}$$

also satisfies (11). Since $H_1$ has a density $\psi$ (with respect to the Lebesgue measure) which satisfies $\psi(H_1) = \psi(-H_1)$ for all $0 < H_1 < \infty$, we can thus assume that the strategy $h^*$ which satisfies (11) is symmetric and has a density.

It follows from Lemma 1 that

$$\tilde{c}(s_{\bar{a}}, h^*) = q,$$

and since (11) holds, it follows that for any search trajectory $S_{\bar{a},u} = \{\bar{a}^{i+u}\}_{i=-\infty}^{\infty}$, $0 \leq u < 2$,

$$\tilde{c}(S_{\bar{a},u}, h^*) = q. \tag{12}$$

We now show that (12) leads to a contradiction. It follows from (12) that for all $0 \leq u < 2$,

$$q = \sum_{n=-\infty}^{\infty} \int_{x_{n-2}}^{x_n} \frac{2 \sum_{j=-\infty}^{n-1} x_j + H}{H} \psi(H) \, dH$$

$$= 1 + 2 \sum_{n=-\infty}^{\infty} x_n \left( \int_{x_{n-1}}^{\infty} \frac{\psi(H)}{H} \, dH + \int_{x_n}^{\infty} \frac{\psi(H)}{H} \, dH \right)$$

$$\stackrel{\text{def}}{=} F(\ldots, x_0, x_1, \ldots, x_n, \ldots),$$

where $x_n = \bar{a}^{n+u}$. Now

$$\frac{1}{2} \frac{\partial F}{\partial x_n} = \int_{x_{n-1}}^{\infty} \frac{\psi(H)}{H} \, dH + \int_{x_n}^{\infty} \frac{\psi(H)}{H} \, dH$$

$$- x_n \frac{\psi(x_n)}{x_n} - x_{n+1} \frac{\psi(x_n)}{x_n}$$

$$= \int_{x_n/\bar{a}} \frac{\psi(H)}{H} \, dH + \int_{x_n}^{\infty} \frac{\psi(H)}{H} \, dH - (1 + \bar{a})\psi(x_n).$$

It follows from (11) and (12) that $\partial F/\partial x_n = 0$ for all $-\infty < n < \infty$ and all $0 \le u < 2$, and since the set $\{a^{n+u}, -\infty < n < \infty, 0 \le u < 2\}$ covers the whole positive axis, it follows that the probability density $\psi(H)$ satisfies for all $\gamma > 0$

$$\left( \int_{\gamma/\bar{a}}^{\infty} \frac{\psi(H)}{H} \, dH + \int_{\gamma}^{\infty} \frac{\psi(H)}{H} \, dH \right) \Big/ \psi(\gamma) = 1 + \bar{a}. \tag{13}$$

We now show that equation (13) implies that $\psi(H) \sim 1/H$ for $H > 0$. Let

$$X(\theta) = \psi(e^{\theta}), \qquad -\infty < \theta < \infty. \tag{14}$$

Then

$$\int_{\gamma/\bar{a}}^{\infty} \frac{\psi(H)}{H} \, dH + \int_{\gamma}^{\infty} \frac{\psi(H)}{H} \, dH$$

$$= \int_{\ln\gamma - \ln\bar{a}}^{\infty} X(\theta) \, d\theta + \int_{\ln\gamma}^{\infty} X(\theta) \, d\theta$$

$$= \int_{-\ln\bar{a}}^{\infty} X(\ln \gamma + \theta) \, d\theta + \int_{0}^{\infty} X(\ln \gamma + \theta) \, d\theta.$$

Let $y = \ln \gamma$. Then (13) and (14) imply that for all $-\infty < y < \infty$

$$\frac{\int_{-\infty}^{\infty} X(y+\theta)\, dA(\theta)}{X(y)} = \frac{\int_{-\ln\bar{a}}^{\infty} X(y+\theta)\, d\theta + \int_{0}^{\infty} X(y+\theta)\, d\theta}{X(y)}$$

$$= 1 + \bar{a}, \tag{15}$$

where $A(\theta)$ is a positive measure with density equal to 2 for $0 \le \theta < \infty$, density equal to 1 for $-\ln \bar{a} \le \theta < 0$, and zero density for $\theta < -\ln \bar{a}$.

We can now use Theorem 6 of Section 6.5. The bilateral Laplace transform $f(b)$ of $A(\theta)$ is finite for all $b < 0$ and satisfies

$$f(b) = \int_{-\ln\bar{a}}^{\infty} e^{b\theta}\, d\theta + \int_{0}^{\infty} e^{b\theta}\, d\theta = -(\bar{a}^{-b} + 1)/b.$$

Thus, the derivative of f satisfies

$$f'(b) = (b\bar{a}^{-b} \ln \bar{a} + \bar{a}^{-b} + 1)/b^2.$$

It follows from (7) that

$$f'(-1) = -\bar{a} \ln \bar{a} + \bar{a} + 1 = 0,$$

so that $f(b)$ attains its minimum at $b = -1$. (The point $b = -1$ is the global minimum because the bilateral Laplace transform is a convex function.)

Since $f(-1) = \bar{a} + 1$, it follows from (15) that

$$\frac{\int_{-\infty}^{\infty} X(y+\theta)\, dA(\theta)}{X(y)} = f(-1)$$

for all $-\infty < y < \infty$, and thus Theorem 6 of Section 6.5 implies that

$$X(\theta) = \alpha e^{-\theta} \qquad \text{for all} \quad -\infty < \theta < \infty,$$

where $\alpha$ is a positive constant. By using (14), it follows that the density $\psi$ of the strategy $h^*$ satisfies for all $0 < H < \infty$

$$\psi(H) = X(\ln H) = \alpha/H,$$

but since $\int_0^\infty 1/H = \infty$, no such density exists. Q.E.D.

Remark 2. It was proven by Beck and Newman (1970) that any hiding strategy h with $E|H| = \lambda$ satisfies

$$\inf_s c(s, h) < q\lambda.$$

This result does not imply Theorem 2. On the other hand, the preceding inequality is easily obtained from Theorem 2; if there did exist a hiding strategy $h^*$ with $E|H| = \lambda$ which satisfies

$$\inf_s c(s, h^*) \geq q\lambda,$$

then the hiding distribution $\tilde{h}$ defined by

$$d\tilde{h}(H) = (|H|/\lambda)\, dh^*(H)$$

would satisfy

$$\inf_s \tilde{c}(s, \tilde{h}) = \inf_s \int \frac{c(s, H)}{\lambda}\, dh^*(H) \geq q,$$

which would contradict Theorem 2.

It is worth noting that the worst possible outcome of using the search strategy $s_{\bar{a}}$ ($\bar{a} \approx 3.6$) is a loss of $1 + 2 \sum_{j=-\infty}^{1} \bar{a}^{-j} \approx 10.9$, while the expected cost of the strategy $s_2$ which uses only minimax trajectories is $1 + 3/(\ln 2) \approx 5.3$. Thus, use of $s_{\bar{a}}$ yields (the minimal) expected cost of 4.6 but risks a maximal cost of 10.9, while use of $s_2$, which yields an expected cost of 5.3, minimizes the maximal cost (which in this case is equal to 9). The expected cost of any search strategy $s_a$ with $2 < a < \bar{a}$ lies between 4.6 and 5.3, while the maximal cost lies between 9 and 10.9. All the strategies $s_a$ with the parameter a lying outside the segment $[2, \bar{a}]$ are dominated by the family $\{s_a : 2 \leq a \leq \bar{a}\}$ with respect to the expected and the maximal cost.

## 7.4 ANOTHER APPROACH TO THE SEARCH ON THE LINE

The discussion presented in Sections 7.2 and 5.1 implies that the optimal pure and mixed strategies presented in Sections 7.2 and 7.3 are still optimal if one uses the capture time as the cost function, with the restriction $E|H| \leq 1$. In order to obtain optimal search strategy in the case that the constraint is $E|H| \leq \lambda$, one has to use the scaling lemma of Chapter 1 with the mapping $\varphi(x) = \lambda x$, but it is readily seen that $\varphi$ maps the optimal search strategy into itself. Thus, the minimax trajectory and the optimal mixed strategy are uniformly optimal for all $\lambda > 0$, so that it is not necessary to have any knowledge about the value of $\lambda$.

We have already noted, in Remark 1 of Section 5.1, that the equivalence between the two approaches holds whenever assumption 1 of Section 5.1 holds. We now describe a search game in a search space which does not satisfy this assumption. Consider the same game described in Section 7.1, except that the search space Q is the half line $-\varepsilon \leq x < \infty$, where $\varepsilon > 0$. If one uses the normalized cost function $\tilde{C}$, then the minimax pure value $\overline{VP}$ would be 9, the same as in Section 7.2, because by Corollary 1 of Section 6.3, any search trajectory S satisfies

$$\overline{\lim_{i\to-\infty}} \frac{\sum_{j=-\infty}^{i+1} x_j}{x_i} \geq \min_{a>0} \sum_{j=-\infty}^{1} a^j = 4.$$

On the other hand, in the case of using the capture time as the cost and restricting the hider to strategies which satisfy $E|H| \leq 1$, if the searcher uses the trajectory $S^*$ which satisfies

$$S^*(t) = -t \qquad \text{for} \quad 0 \leq t \leq \varepsilon,$$

and

$$S^*(t) = t - 2\varepsilon \qquad \text{for} \quad \varepsilon < t,$$

then, for any H,

$$C(S^*, H) \leq |H| + 2\varepsilon.$$

Thus, for any admissible search strategy,

$$c(S^*, h) \leq 1 + 2\varepsilon,$$

so that the pure value is much less than 9.

## 7.5 SEARCH FOR A MOVING HIDER

This problem is an extension of the search on the real line to the case of a moving hider. At time $t = 0$, the hider is located at a point $-\infty < H(0) < \infty$ on the real line, and at any $t > 0$, the hider moves away from the origin with velocity $w < 1$. All the other rules of the game are the same as those considered so far.

For any search trajectory $S = \{x_i\}_{i=-\infty}^{\infty}$ the loss of the searcher $\tilde{C}(S, H)$ is equal to the distance $C(S, H)$ traveled by the searcher (until he meets the hider) divided by $|H(0)|$.

We are interested in the minimax pure strategy of the searcher; thus, we want to find

$$\inf_S \sup_H \tilde{C}(S, H)/|H(0)|.$$

At first, note that if the hider was not found at stage number $i - 2$, then in order to have a chance to find him at stage number i, the searcher has to use a step length $x_i$ such that $x_i > d$, where d satisfies $d = x_{i-2} + w(x_{i-2} + 2x_{i-1} + d)$, or $(1 - w)d = (1 + w)x_{i-2} + 2wx_{i-1}$.

Thus, $x_i$ satisfies

$$(1 - w)x_i > (1 + w)x_{i-2} + 2wx_{i-1}$$

or

$$(1 - w)(x_i - x_{i-2}) > 2w(x_{i-2} + x_{i-1}).$$

It follows from the preceding inequality that for every i,

$$(1 - w)x_i > 2w \sum_{-\infty}^{i-1} x_j. \tag{16}$$

Because of this we need only consider trajectories satisfying (16). Suppose that this condition holds and that at stage i the searcher is at a distance $x_i$ and the hider a distance of $x_i + \varepsilon$ from the origin. In this case,

$$|H(0)| + w\left(2 \sum_{-\infty}^{i-1} x_j + x_i\right) = x_i + \varepsilon. \tag{17}$$

Suppose also that the hider has not been found before stage i. One can show that if $\varepsilon$ is a sufficiently small positive number, then the hider will be found at stage i + 2. Assume this holds for some integer i and that the hider is found in stage i + 2 at a distance d from the origin (see Figure 1).

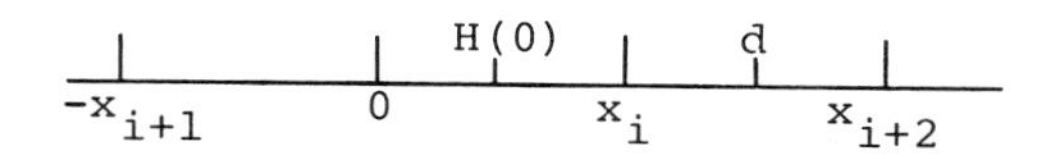


FIGURE 1

In this case, the following relation holds:

$$C(S, H) = 2 \sum_{-\infty}^{i+1} x_j + d, \tag{18}$$

where

$$d = |H(0)| + wC(S, H). \tag{19}$$

Using (18) and (19), we obtain

$$C(S, H) = 2 \sum_{-\infty}^{i+1} x_j + |H(0)| + wC(S, H),$$

so that

$$C(S, H) = \frac{1}{1 - w} |H(0)| + \frac{2}{1 - w} \sum_{-\infty}^{i+1} x_j.$$

Using (17) with $\varepsilon = 0$, we obtain

$$\frac{C(S, H)}{|H(0)|} = \frac{1}{1 - w} + \frac{2}{1 - w} \frac{\sum_{-\infty}^{i+1} x_j}{(1 - w)x_i - 2w \sum_{-\infty}^{i-1} x_j}. \tag{20}$$

Thus, we are interested in

$$\inf_{S} \sup_{-\infty<i<\infty} \frac{\sum_{-\infty}^{i+1} x_j}{(1 - w)x_i - 2w \sum_{-\infty}^{i-1} x_j} \tag{21}$$

for sequences $S = \{x_i : x_i > 0, -\infty < i < \infty\}$ satisfying condition (16).

In order to solve (21), let us denote

$$z_i = \sum_{-\infty}^{i} x_j \tag{22}$$

and

$$y_i = x_i - \frac{2w}{1 - w} \sum_{-\infty}^{i-1} x_j > 0 \qquad \text{(using (16))}. \tag{23}$$

Then

$$y_i = (z_i - z_{i-1}) - \frac{2w}{1 - w} z_{i-1} = z_i - \frac{1 + w}{1 - w} z_{i-1}.$$

In summation, we obtain

$$z_i = \sum_{j=-\infty}^{0} \left(\frac{1 + w}{1 - w}\right)^{-j} y_{i+j}.$$

The preceding equality together with (22) and (23) implies that (21) is equivalent to

$$\frac{1}{1 - w} \inf_{Y} \sup_{-\infty < i < \infty} \sum_{j=-\infty}^{0} \left(\frac{1 - w}{1 + w}\right)^{j} y_{i+j+1} \Big/ y_i, \tag{24}$$

where $Y = \{y_i, -\infty < i < \infty\}$ is any positive sequence.

Problem (24) can be solved by Theorem 7 of Section 6.5; we have to find

$$\frac{1}{1 - w} \inf_{a>0} \sum_{j=-\infty}^{0} \left(\frac{1 - w}{1 + w}\right)^{j} a^{j+1}$$

$$= \frac{1}{1 - w} \inf_{a>0} a \Big/ \left(1 - \frac{1 + w}{a(1 - w)}\right),$$

which is equal to

$$4(1 + w)/(1 - w)^2, \tag{25}$$

the optimal a being

$$\bar{a} = 2(1 + w)/(1 - w).$$

Thus the optimal sequence Y in (24) is $y_i = \alpha\bar{a}^i$, where $\alpha$ is any constant.

It follows from (23) that $x_i = \alpha\bar{a}^i$ is the optimal solution for (21).

Thus, it follows from (20) and (25) that for any search trajectory S, there exists a point H(0) so that

$$\frac{C(S, H)}{|H(0)|} \geqq \frac{1}{1 - w} + \frac{8(1 + w)}{(1 - w)^3}. \tag{26}$$

On the other hand, if we use

$$x_i = \alpha(2(1 + w)/(1 - w))^i, \qquad -\infty < i < \infty, \tag{27}$$

and $\alpha$ is any positive constant, then

$$(1 - w)x_i = \alpha(1 - w)\left(\frac{2(1 + w)}{1 - w}\right)^i$$

$$> \alpha \frac{2w}{1 + 3w}(1 - w)\left(\frac{2(1 + w)}{1 - w}\right)^i = 2w \sum_{-\infty}^{i-1} x_j, \tag{28}$$

so that (16) is satisfied.

We now proceed to establish the fact that $S^*$ defined by (27) is the minimax trajectory.

At first note that for any H, $C(S^*, H) < \infty$. This is due to the fact that (27) implies

$$x_i - w\left(2 \sum_{-\infty}^{i-1} x_j + x_i\right) \sim \left(\frac{2(1 + w)}{1 - w}\right)^i \to \infty \qquad \text{as} \quad i \to \infty,$$

so that for any H, if i is large enough, then

$$x_i > |H(0)| + w\left(2 \sum_{-\infty}^{i-1} x_j + x_i\right).$$

Thus, the hider can always be found.

In order to complete the proof, we must show (see (26)) that if $S^*$ is defined by (27), then for any H,

$$\frac{C(S^*, H)}{|H(0)|} \leqq \frac{1}{1 - w} + \frac{8(1 + w)}{(1 - w)^3}. \tag{29}$$

We use the following argument. Assume that in stage i + 2 the hider is found at a distance d from the origin. Then relations (18) and (19) hold while (17) implies that

$$|H(0)| > (1 - w)x_i - 2w \sum_{-\infty}^{i-1} x_j.$$

Thus, instead of (20), we obtain

$$\frac{C(S^*, H)}{|H(0)|} \leqq \frac{1}{1 - w} + \frac{2}{1 - w} \frac{\sum_{-\infty}^{i+1} x_j}{(1 - w)x_i - 2w \sum_{-\infty}^{i-1} x_j}, \tag{30}$$

but we have previously shown that if $S^*$ is defined by (27), then (for all i) the right side of (30) is equal to $(1/(1 - w)) + 8(1 + w)/(1 - w)^3$, so that (29) is established.

Note that if $w = 0$, then the minimax search trajectory is $\{\alpha 2^i, -\infty < i < \infty\}$, as has been established in Section 7.2.

## 7.6 SEARCH WITH PROBABILITY OF DETECTION LESS THAN 1

### Presentation of the Problem

An object is located at a point H on the real line. Suppose that the probability that the object will be discovered when the searcher passes it the i-th time is $P_i$, where

$$\sum_{i=1}^{\infty} P_i = 1. \tag{31}$$

The loss function $\widetilde{C}(S, H)$ is defined as the expected distance traveled by the searcher (until the object has been found) divided by $|H|$.

We shall now find the minimax periodic monotonic trajectory. In this case, if

$x_I \leq |H| \leq x_{I+2}$,

then

$$\widetilde{C}(S, H) = \frac{1}{|H|}\left[P_1\left(|H| + 2\sum_{-\infty}^{I+1} x_j\right) + P_2\left(-|H| + 2\sum_{-\infty}^{I+2} x_j\right)\right.$$

$$+ p_3\left(|H| + 2\sum_{-\infty}^{I+3} x_j\right) + p_4\left(-|H| + 2\sum_{-\infty}^{I+4} x_j\right)$$

$$\left. + \ \ldots \right]$$

$$= W + \frac{2}{|H|}\left[\sum_{j=-\infty}^{I+1} x_j + \sum_{j=2}^{\infty} q_j x_{I+j}\right], \tag{32}$$

where

$$q_j = \sum_{i=j}^{\infty} p_i \tag{33}$$

and

$$W = \sum_{j=0}^{\infty} p_{2j+1} - \sum_{j=1}^{\infty} p_{2j}. \tag{34}$$

Obviously,

$$v(S) = W + 2 \sup_{-\infty<I<\infty} \frac{1}{x_I}\left[\sum_{j=-\infty}^{I+1} x_j + \sum_{j=2}^{\infty} q_j x_{I+j}\right]. \tag{35}$$

It follows from Corollary 1 of Section 6.3 that

$$v(S) \geq W + 2 \inf_{0<a<\infty}\left[\sum_{j=-\infty}^{1} a^j + \sum_{j=2}^{\infty} q_j a^j\right]. \tag{36}$$

Thus, the expression given by the right-hand side of (36) is the value that can be obtained by using a periodic monotonic trajectory.

In order to obtain this value, one would have to use the trajectory $x_i = \alpha\bar{a}^i$, for $-\infty < i < \infty$, where $\alpha$ is a constant and $\bar{a}$ is the value of the a which minimizes the right-hand side of (36). In this trajectory, the positive and the negative rays are visited alternatively.

On the other hand, contrary to the preceding problems, the optimal trajectory need not be periodic and monotonic, as will be shown by the following example.

## Search with a Delay

Let $P_3 = 1$ and $P_j = 0$ for $j \neq 3$. It follows from (36) that the value than can be obtained by using periodic monotonic trajectories is equal to

$$1 + 2 \inf_{0<a<\infty} \left[ \sum_{j=-\infty}^{1} a^j + a^2 + a^3 \right] \approx 20.0. \tag{37}$$

On the other hand, a smaller value can be obtained by the following trajectory. Let

$$\cdots < y_{2i+1} < y_{2i-1} < \cdots < 0 < \cdots$$
$$< y_{2i-2} < y_{2i} < y_{2i+2} < \cdots .$$

The trajectory Y is defined as follows.

In stage i of Y we travel $0 \to y_i$, $y_i \to y_{i-2}$, $y_{i-2} \to y_i$, and $y_i \to 0$. The distance traveled at stage i is equal to

$$2|y_i| + 2(|y_i| - |y_{i-2}|) = 4|y_i| - 2|y_{i-2}|.$$

If $y_I < |H| < y_{I+2}$ (or $y_{I+2} < -|H| < y_I$), then

$$\tilde{C}(Y, H) = \frac{1}{|H|}\left[ \sum_{i=-\infty}^{I+1} (4|y_i| - 2|y_{i-2}|) + |y_{I+2}| + |y_{I+2}| - |y_I| + |H| - |y_I| \right]$$
$$= 1 + \frac{2}{|H|}\left[ \sum_{i=-\infty}^{I+2} |y_i| + |y_{I+1}| \right]. \tag{38}$$

Thus

$$v(Y) = 1 + 2 \sup_{-\infty<I<\infty} \left|\frac{1}{y_I}\right| \left[ \sum_{-\infty}^{I+2} |y_i| + |y_{I+1}| \right]. \tag{39}$$

Let $\bar{a}$ be the a that minimizes

$$\sum_{j=-\infty}^{2} a^j + a \qquad (\bar{a} \approx 1.45).$$

If we choose $y_i = \alpha(-1)(-\bar{a})^i$, then $v(Y) = 17.2$ which is less than the value obtained in (37).

It follows from the previous example that in some of the problems with probability of detection less than 1, the optimal trajectory does not have the simple properties of monotonicity and periodicity. We now present the following open problem.

Geometric Detection Probability

Assume that the searcher detects the hider with probability p each time he passes the point occupied by the hider. In other words, the probability of capture the i-th time the searcher visits the point H is $P_i = (1 - p)^{i-1}p$.

It is easily verified that the right-hand side of (36), which is the pure value obtained by monotonic and periodic trajectories, is

$$\overline{VP}_1 = (p/(2 - p)) + (8/p). \tag{40}$$

(The optimal parameter of the geometric trajectories is $2/(2 - p)$.)

Thus, if the minimal trajectory is monotonic and periodic, then $\overline{VP}_1$ given by (40) is the pure value of the game. However, it is still an open problem whether the minimax trajectory is indeed monotonic and periodic. A similar difficulty is associated with the optimal mixed strategy.

Chapter 8

# Application Of The Minimax Theorems To Some Other Search Problems

## 8.1 SEARCH ON M RAYS

In this section we are concerned with the problem of finding an immobile hider in the set Q which consists of M (M > 1) unbounded rays radiating from the origin O. This problem is an extension of the search on the real line, where M = 2, and the results that we obtain are indeed of a similar nature. We now present a detailed description of the problem: A pure strategy of the hider is given by

$$H = (|H|, m), \tag{1}$$

where $1 \leq m \leq M$ is the ray number and $|H| > 0$ is the distance from the origin (Figure 1).

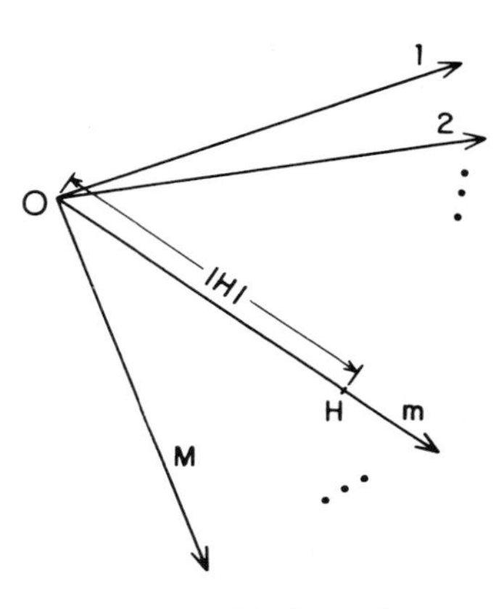


FIGURE 1

It follows from the discussion presented in Section 7.1 for the search on the line (M = 2) that any admissible trajectory S of the searcher can be described by an infinite number of pairs

$$S = \{(x_i, N_i)\} \tag{2}$$

where $-\infty < i < \infty$, $0 < x_i$, and $1 \le N_i \le M$. At the i-th stage of the trajectory S, the searcher starts from the origin, walks a distance $x_i$ (or $|H|$ if the hider is to be found at this stage) along ray $N_i$, and returns to the origin O.

The capture time C(S, H) is given by

$$C(S, H) = 2 \sum_{i=-\infty}^{i_H - 1} x_i + |H|, \tag{3}$$

where $i_H$ is the number of the stage during which the hider is discovered.

We shall use the normalized cost function $\tilde{C}(S, H)$, discussed in Section 5.1 and used in Chapter 7. Thus the cost is the capture time divided by the distance of the hiding point form the origin,

$$\tilde{C}(S, H) = C(S, H)/|H|. \tag{4}$$

As usual, we denote the value of the search trajectory S by

$$v(S) = \sup_H \tilde{C}(S, H). \tag{5}$$

In order to find the minimax search trajectory, we first establish several properties of the search trajectories. At first, we introduce a representation of the search trajectories which can be handled more conveniently than (2)

$$\bar{S} = \{(x_i, J_i)\}_{i=-\infty}^{\infty}, \tag{6}$$

where $x_i$ is the distance traveled along the chosen ray at the i-th stage, and $J_i$ is the minimal stage number greater than i, in which the same ray is visited again. The function $J_i$ should satisfy the following three requirements.

For every i, $i < J_i < \infty$. (7)

For every i, there exists an $l < i$ such that

$J_l = i$ ($l$ will be denoted by $J_i^{-1}$). (8)

For every integer $l$, the set

$G_l = \{(i : i \leq l \text{ and } J_k > l\}$

contains M members. (9)

For example, if M = 4 and the trajectory S satisfies (see (2))

$\ldots, N_{-3} = 3, N_{-2} = 4, N_{-1} = 2, N_0 = 1,$

$N_1 = 4, N_2 = 2, N_3 = 3, N_4 = 1, \ldots,$

then $J_{-3} = 3$, because the next visit to ray number 3, which is visited when i = -3, occurs when i = 3; similarly, $J_{-2} = 1$, $J_{-1} = 2$, $J_0 = 4, \ldots,$ etc. $J_1^{-1} = -2$, because the last visit to ray number 4 before i = 1 occurs when i = -2; similarly, $J_2^{-1} = -1$, $J_3^{-1} = -3$, $J_4^{-1} = 0, \ldots,$ etc., and $G_1 = \{-3, -1, 0, 1\}$ because ray number 3 is visited when $\underline{i = -3} \leq 1$ and after that when $i = 3 > 1$, ray number 2 is visited when $\underline{i = -1} \leq 1$ and when $i = 2 > 1$, ray number 1 is visited at $\underline{i = 0} \leq 1$ and $i = 4 > 1$, and ray number 4 is visited at $\underline{i = 1} \leq 1$. Similarly,

$G_2 = \{-3, 0, 1, 2\}, \quad G_3 = \{0, 1, 2, 3\}, \quad \text{etc.}$

It is obvious that for each S, the set of trajectories defined by (6) contains M! trajectories which can be generated from one strategy by changing the enumeration of the M rays. Thus, for each $S \in \overline{S}$, v(S) is the same, and so $v(\overline{S})$ can be defined by this value.

It is also obvious that if the function $J_i$ does not satisfy any one of the requirements (7)-(9), then either after or before a certain stage the searcher will not visit a certain ray, so that the value of such a trajectory is infinite.

For example, if $M = 3$ and the trajectory S satisfies (see (2))

$$N_{2i} = 1 \quad \text{and} \quad N_{2i+1} = 2 \quad \text{for} \quad -\infty < i < \infty,$$

then ray number 3 is not visited and indeed, for all $l$, $G_l = \{l - 1, l\}$ contains only two members.

On the other hand, if the trajectory S satisfies (7)-(9), then it follows from (7) and (8) that for any visit on a certain ray there corresponds a following visit and also a preceding visit on this ray. In addition, we note that if at least one ray is not visited, then $G_l$ contains less than M members for all $l$. Thus, it follows from (9) that all the rays are visited and from (7) and (8) that for each of them, there is no first visit and no last visit.

It follows from dominance considerations that we may deal only with trajectories having the monotonicity property defined as

$$\text{for every } i, \quad x_{J_i} > x_i. \tag{10}$$

Obviously, we may also assume that

$$V(\overline{S}) < \infty. \tag{11}$$

A trajectory $\overline{S}$ satisfying conditions (7)-(11) will be called <u>an admissible trajectory</u>. We shall show that for all M, there exist admissible trajectories.

For each admissible trajectory $\overline{S}$,

$$v(\overline{S}) = \sup_H \frac{C(\overline{S}, H)}{|H|} = 1 + 2 \sup_{-\infty<i<\infty} b_i, \tag{12}$$

where

$$b_i = \frac{\sum_{j=-\infty}^{J_i - 1} x_j}{x_i}. \tag{13}$$

This follows from the fact that against a known trajectory S, the "best" point for the hider belongs to the set $\{(x_i + 0, N_i)\}$ (see (2)).

A trajectory X satisfying

$$\text{for each } i, \qquad J_i = i + M, \tag{14}$$

will be called a <u>periodic trajectory</u>. If $\bar{S}$ is a periodic (and monotonic) trajectory, then

$$v(\bar{S}) = 1 + 2 \sup_{-\infty<i<\infty} \frac{\sum_{j=-\infty}^{i+M-1} x_j}{x_i}. \tag{15}$$

It follows from Corollary 1 of Section 6.3 that

$$\sup_{-\infty<i<\infty} \frac{\sum_{j=-\infty}^{i+M-1} x_j}{x_i} \geq \inf_{0<a<\infty} \sum_{j=-\infty}^{M-1} a^j = \frac{M^M}{(M - 1)^{M-1}}. \tag{16}$$

Furthermore, it follows from Theorem 7 of Section 6.5 that if

$$\sup_{-\infty<i<\infty} \frac{\sum_{j=-\infty}^{i+M-1} x_j}{x_i} = \frac{M^M}{(M - 1)^{M-1}},$$

then

$$x_i = \alpha\left(\frac{M}{M - 1}\right)^i, \tag{17}$$

where $\alpha$ is a positive constant.

Hence, the trajectory $\{(x_i, J_i)\}$, where $x_i$ is given by (17) and $J_i = i + M$, which is obviously an admissible trajectory, is optimal among the periodic trajectories. We shall show that it is the best (pure) strategy among all the possible trajectories.

To this end, we present the following lemmas.

LEMMA 1. Every admissible trajectory satisfies

$$\lim_{i\to-\infty} x_i = 0 \tag{18}$$

and

$$\lim_{i\to+\infty} x_i = \infty. \tag{19}$$

These results are obvious.

LEMMA 2. For each trajectory $\bar{S} = (x_i, J_i)$, there exists a trajectory $S^1 = \{(x_i^1, J_i^1)\}$ satisfying, for every $-\infty < i < \infty$,

$$x_{i+1}^1 \geq x_i^1 \tag{20}$$

and

$$\sup_{-\infty<i<\infty} b_i^1 \leq \sup_{-\infty<i<\infty} b_i, \tag{21}$$

where $b_i$ is defined by (13) and

$$b_i^1 = \frac{\sum_{j=-\infty}^{J_i^1-1} x_j^1}{x_i^1}. \tag{22}$$

Furthermore, if there exists an $i_0$ satisfying $x_{i_0+1} < x_{i_0}$, then there exists a $j_0$ satisfying $b_{j_0}^1 < b_{j_0}$.

Proof. Assume that there exists an $i_0$ such that

$$x_{i_0+1} < x_{i_0}. \tag{23}$$

Let us introduce a set of operators $\{A_i\}$, $-\infty < i < \infty$, where $A_i(\bar{S})$ is defined as

$$A_i(\bar{S}) = \{(A_i x_l, A_i J_l)\}, \qquad -\infty < l < \infty, \tag{24}$$

where

$$A_i x_l = x_l \quad \text{and} \quad A_i J_l = A_i J_l \quad \text{for} \quad l \neq i,\ i+1$$

$$A_i x_i = x_{i+1} \quad \text{and} \quad A_i J_i = J_{i+1}$$

$$A_i x_{i+1} = x_i \quad \text{and} \quad A_i J_{i+1} = J_i.$$

Thus, if the operator $A_i$ is applied to a given trajectory $\overline{S}$, then the trajectory is unaffected before the i-th stage. From the i-th stage on, the role of the rays visited at stages number i and i + 1 is interchanged.

We define

$$A_i b_l \overset{\text{def}}{=} \frac{\sum_{j=-\infty}^{A_i J_l - 1} A_i x_j}{A_i x_l}. \tag{25}$$

It follows from (23) and (24) that for each $l \neq J^{-1}_{i_0+1}$, $i_0$, or $i_0 + 1$; $A_{i_0} b_l = b_l$, while

$$\left.\begin{aligned} A_{i_0} b_{i_0} &= b_{i_0+1}, \\ A_{i_0} b_{i_0+1} &= b_{i_0}, \\ A_{i_0} b_l &< b_l \qquad \text{for} \quad l = J^{-1}_{i_0+1}. \end{aligned}\right\} \tag{26}$$

Lemma 1 implies that starting with $\overline{S}$ and applying operators from the set $\{A_i\}$ a finite number of times, it is possible to obtain a trajectory $\overline{S}^{(2)} = \{(x_i^{(2)}, J_i^{(2)})\}$ satisfying the following condition.

There exist $i_1^{(2)} < i_2^{(2)}$ such that

$$x_i^{(2)} \leq 1/2 \qquad \text{for} \quad i \leq i_1^{(2)},$$

$$x_i^{(2)} \geq 2 \qquad \text{for} \quad i \geq i_2^{(2)},$$

$$1/2 \le x_i^{(2)} \le 2 \quad \text{for} \quad i_1^{(2)} < i < i_2^{(2)},$$

$$x_i^{(2)} \le x_l^{(2)} \quad \text{if} \quad i_1^{(2)} < i < l < i_2^{(2)}. \tag{27}$$

Continuing in a similar manner, we can define recursively a sequence of trajectories $\bar{S}^{(n)}$ satisfying

$$\bar{S}^{(n+1)} = A_{n_{k_n}}(A_{n_{k_n-1}}(\ldots(A_{n_1}(\bar{S}^{(n)}))\ldots)), \tag{28}$$

where $n_j \le i_1^{(n)}$ or $n_j \ge i_2^{(n)}$, and there exist $i_1^{(n+1)} < i_2^{(n+1)}$ such that

$$\begin{aligned} x_i^{(n+1)} &\le 1/(n+1) \\ &\quad \text{for } i \le i_1^{(n+1)}, \\ x_i^{(n+1)} &\ge n+1 \\ &\quad \text{for } i \ge i_2^{(n+1)}, \\ 1/(n+1) \le x_i^{(n+1)} &< n+1 \\ &\quad \text{for } i_1^{(n+1)} < i < i_2^{(n+1)}, \\ x_i^{(n+1)} &\le x_l^{(n+1)} \\ &\quad \text{if } i_1^{(n+1)} < i < l < i_2^{(n+1)}. \end{aligned} \tag{29}$$

It follows from (24) and (28) that if $i_1^{(n)} < i < i_2^{(n)}$, then for all $N > n$,

$$x_i^{(N)} = x_i^{(n)}, \qquad J_i^{(N)} = J_i^{(n)}. \tag{30}$$

Hence

$$x_i^1 = \lim_{n\to\infty} x_i^{(n)} \quad \text{and} \quad J_i^1 = \lim_{n\to\infty} J_i^{(n)}$$

exist. If the trajectory $S^1$ is defined as $S^1 = \{(x_i^1, J_i^1)\}$, then it follows from (26) and (29) that $S^1$ satisfies the inequalities (20) and (21). Q.E.D.

LEMMA 3. Let

$$d_i = \frac{\sum_{j=-\infty}^{i+M-1} x_j}{x_i}, \tag{31}$$

$$D_i = \left(\prod_{l=i}^{i+M-1} d_l\right)^{1/M}, \tag{32}$$

and

$$q = \inf_{0<a<\infty} \sum_{j=-\infty}^{M-1} a^j = \frac{M^M}{(M-1)^{M-1}}. \tag{33}$$

If $D_i \le q$ holds for each $i$, then $D_i = q$.

Proof. Define

$$y_i = \left(\prod_{l=i}^{i+M-1} x_l\right)^{1/M}, \tag{34}$$

$$E_i = \frac{\sum_{j=-\infty}^{i+M-1} y_j}{y_i}. \tag{35}$$

Using Hölder's inequality (see Hardy et al., 1952), we obtain

$$D_i \ge \frac{\sum_{j=-\infty}^{i+M-1} \left(\prod_{l=j}^{j+M-1} x_l\right)^{1/M}}{\left(\prod_{l=i}^{i+M-1} x_l\right)^{1/M}} = E_i. \tag{36}$$

Since $E_i \le D_i \le q$, it follows from (36) and Theorem 7 of Section 6.5, that $E_i = q$ and consequently $D_i = q$ for all $i$. Q.E.D.

It is now possible to show that the periodic trajectory defined by (17) is the minimax pure strategy for the searcher, and that the pure value of the searcher, $\overline{VP}$, is equal to

$$1 + 2 \frac{M^M}{(M - 1)^{M-1}}.$$

THEOREM 1. Let $\bar{S}$ be a search trajectory and let $b_i$ and $q$ be defined by (13) and (33). If for all $-\infty < i < \infty$,

$$b_i \leqq q, \tag{37}$$

then for all $i$

(a) $b_i = q$,

(b) $J_i = i + M$, and

(c) $x_i = \alpha(M/(M - 1))^i$, where $\alpha$ is any positive constant.

Proof. The proof is given in two parts:

1. First assume that

$$x_i \leqq x_{i+1} \qquad \text{for every} \quad -\infty < i < \infty. \tag{38}$$

Since $\bar{S}$ is an admissible strategy, we can use condition (9) together with (37) and deduce that for every $-\infty < l < \infty$,

$$q \geqq \Big(\prod_{i \in G_l} b_i\Big)^{1/M}$$

$$\geqq \left(\prod_{i=l-M+1}^{l} \frac{\sum_{j=-\infty}^{i+M-1} x_j}{x_i}\right)^{1/M} \qquad \text{(by (38))}$$

$$= D_{l-M+1}, \tag{39}$$

$D_{l-M+1}$ being defined by (32). Using Lemma 3, we deduce that for every $l$, $D_l = q$. Finally, we use (37) and (39) and establish (a).

Assumption (38) together with (a) implies that

$$i < l \to J_i < J_l. \tag{40}$$

This obviously implies that the set $G_l$ defined by (9) consists of the integers $l - M + 1$, $l - M + 2, \ldots, l$, and that $G_{l-1} = \{l - M, l - M + 1, \ldots, l - 1\}$.

Thus for every $l$, $J_{l-M} = l$, so that (b) is established.

In order to prove (c), it remains to note that proposition (b) implies that $b_i = \sum_{j=-\infty}^{i+M-1} x_j/x_i$, so that (c) follows from (37) and Theorem 7 of Section 6.5. This completes the proof under assumption (38).

2. If there exists an integer $i_0$ satisfying

$$x_{i_0+1} < x_{i_0}, \tag{41}$$

then we can use Lemma 2 and obtain a trajectory $S^1$ satisfying, for all $-\infty < i < \infty$,

$$x_i^1 \leqq x_{i+1}^1, \qquad b_i^1 \leqq q, \qquad b_{i_0}^1 < q.$$

However, the existence of this trajectory contradicts part 1. Therefore inequality (41) cannot hold. Q.E.D.

We now present, as an example, the optimal search strategy for the case where $M = 3$. Using Theorem 1, we see that the optimal strategy has the form $\bar{S} = \{(\alpha(3/2)^i, i + 3)\}$, where $\alpha$ is any positive constant.

If we choose $\alpha = 1$, then the stages $i = 0$ to $i = 4$ (for example) are carried out by going a distance of 1 along the first ray (and returning to the origin), then a distance of 3/2 along the second ray, then a distance of $3^2/2^2$ along the third ray and then a distance of $3^3/2^3$ along the first ray, etc. The value of this game is $1 + 2(3^3/2^2) = 14.5$.

We have previously discussed the fact that the optimal trajectory has no first step. In order to apply it to a real world situation, one would have to assume that the object cannot be hidden closer than $\varepsilon$ from the origin. In this case we can modify the optimal trajectory and define it to be

$$x_i = \varepsilon\left(\frac{M}{M-1}\right)^i, \qquad 0 \leqq i \leqq \infty,$$

and $J_i = i + M$ as before.

Thus, in this trajectory there is a first step which is to travel a distance $\varepsilon$ on the first ray and return to the origin. The second step is to travel a distance of $\varepsilon(M/M-1)$ on the second ray and return to the origin, etc.

Remark 1. It is very reasonable that the results presented in Section 7.3 can be extended to M rays. The optimal (mixed) strategy should be periodic and monotonic with $x_i = \bar{a}^{i+u}$, where u is uniformly distributed in [0, M) and $\bar{a}$ minimizes

$$\tilde{c}(s_a, H) = 1 + \frac{2}{|H|} E\left(\sum_{j=-\infty}^{i+M-1} a^{j+u} \,\middle|\, a^{i+u} < |H| \le a^{i+M+u}\right)$$

$$= 1 + \frac{2(a^M - 1)}{M(a-1)\ln a}.$$

However, the proof that $s_{\bar{a}}$ is indeed optimal may involve some technical complications.

## 8.2 SEARCH FOR A POINT IN THE PLANE

In this section, we use the results presented in Section 6.4 in order to find the minimax search trajectory for the following search problem in the plane. The (immobile) hider is located at a point in the plane, $H = (|H|, \beta)$, where $|H|$ is the distance from the origin O and $\beta$ is the angle of OH. The searcher chooses a trajectory starting from the origin. He discovers the hider at the moment when H is <u>covered by the area swept by the radius vector of his trajectory</u>.

We shall consider search trajectories $S(\gamma)$ in which the angle $-\infty < \gamma < \infty$, of the radius vector $X(\gamma)$, is always increasing, and X also satisfies $X(\gamma + 2\pi) \geq X(\gamma)$ for all $\gamma$. In addition, we assume that $X'(\gamma)$ is piecewise continuous and bounded on any finite interval.

The capture time $C(S, H)$ is the length of the trajectory traveled by the searcher until the hider is discovered. Thus,

$$C(S, H) = \int_{-\infty}^{\gamma(S,H)} \sqrt{X^2(\gamma) + X'^2(\gamma)}\, d\gamma,$$

the upper limit $\gamma(S, H)$ of the integral being equal to the value of $\gamma$ at which the hider is to be discovered. The part of the trajectory used until the hider is found, is illustrated in Figure 2.

We shall use the normalized cost function

$$\tilde{C}(S, H) = C(S, H)/|H|. \tag{42}$$

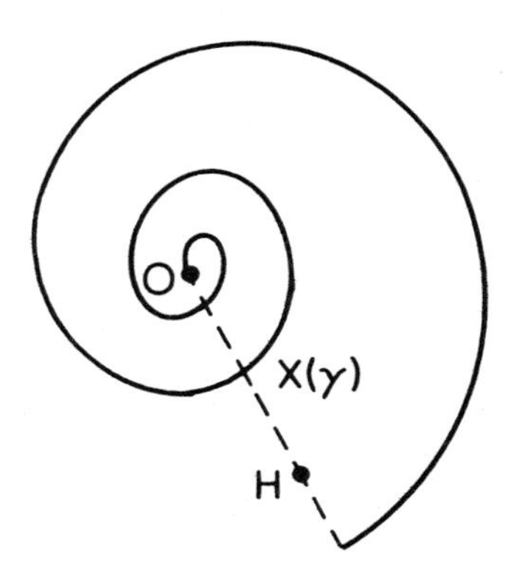


FIGURE 2

Our aim is to find a trajectory $S(\gamma)$ which minimizes $\sup_H \tilde{C}(S, H)$. Thus we are interested in

$$\inf_X \sup_H \left( \int_{-\infty}^{\gamma(S,H)} \sqrt{X^2(\gamma) + X'^2(\gamma)}\, d\gamma \right) / |H|.$$

Since $|H| > X(\gamma(S, H) - 2\pi)$, we note that the problem is equivalent to

$$\inf_X \sup_{-\infty<\tau<\infty} \int_{-\infty}^{\tau} \sqrt{X^2(\gamma) + X'^2(\gamma)}\, d\gamma / X(\tau - 2\pi). \tag{43}$$

Problem (43) can be solved as follows. Define

$$F_k(X) = \int_{-k}^{0} \sqrt{X^2(\gamma) + X'^2(\gamma)}\, d\gamma / X(-2\pi), \qquad k > 2\pi.$$

In order to use Theorem 5 of Section 6.4, we have to show that the functionals $F_k$ satisfy the condition required by that theorem. At first we note that $F_k(\alpha X) = F_k(X)$ for any $\alpha > 0$ and that $F_k(X)$ is continuous with respect to the metric $d(X, Y) = \sup_{-k\le\gamma\le 0}(|X(\gamma) - Y(\gamma)| + |X'(\gamma) - Y'(\gamma)|)$. We now show that

$$F_k(X + Y) \le \max(F_k(X), F_k(Y)).$$

Assume that $F_k(X) \le \alpha$ and $F_k(Y) \le \alpha$ and let $W = X + Y$. Then

$$\int_{-k}^{0} \left( \sqrt{X^2(\gamma) + X'^2(\gamma)} + \sqrt{Y^2(\gamma) + Y'^2(\gamma)} \right) d\gamma$$

$$\le \alpha(X(-2\pi) + Y(-2\pi)) = \alpha W(-2\pi).$$

On the other hand, it follows from Minkowski's inequality (see Hardy et al., 1952) that

$$\sqrt{W^2(\gamma) + W'^2(\gamma)} = \sqrt{(X(\gamma) + Y(\gamma))^2 + (X'(\gamma) + Y'(\gamma))^2}$$
$$\le \sqrt{X^2(\gamma) + X'^2(\gamma)} + \sqrt{Y^2(\gamma) + Y'^2(\gamma)},$$

so that $F_k(W) \le \alpha$.

In order to use Theorem 5 of Section 6.4, we also have to show that $F_k$ satisfies (41) and (42) of that section. Specifically, we now show that if $b_x = \pm\infty$ (see equation (40), Section 6.4), then $\overline{\lim}_{R\to\infty} F_k(X^{+R}) = \infty$. This follows from the fact that

$$F_k(X) \geq \int_{-k}^{0} X(\gamma)\, d\gamma / X(-2\pi),$$

so that the argument used in the proof of Corollary 2 of Section 6.4 is also valid here.

From the preceding discussion, it follows that $F_k(X)$ satisfies the conditions of Theorem 5 of Section 6.4; hence

$$\overline{\lim_{\tau\to\infty}} \frac{\int_{-\infty}^{\tau} \sqrt{X^2(\gamma) + X'^2(\gamma)}\, d\gamma}{X(\tau - 2\pi)} \geq \inf_{-\infty<b<\infty} \int_{-\infty}^{2\pi} \sqrt{e^{2b\gamma} + b^2 e^{2b\gamma}}\, d\gamma$$

$$= \inf_{0<b<\infty} e^{2\pi b} \sqrt{1 + (1/b^2)}. \qquad (44)$$

Let $\bar{b}$ be the value of b which minimizes (44). Then the minimax search trajectory is $X(\gamma) = \alpha e^{\bar{b}\gamma}$, where $\alpha$ is a positive constant. (Note that $\alpha = \alpha_1$ and $\alpha = \alpha_1 e^{2\pi j\bar{b}}$, where j is any integer, represent the same trajectory.) The solution is determined up to a multiplicative constant because $F(\alpha X) = F(X)$ for all positive constants $\alpha$.

It should be noted that, similar to the previous problem $S(\gamma)$ starts with small "oscillations." Thus, in order to use $S(\gamma)$, one would have to assume that the object is outside a circle of radius $\varepsilon$ around the origin (Figure 3). The optimal trajectory can now be modified by initially moving a distance $\varepsilon$ along a radius of this circle, in the direction where $\gamma = 0$, and then proceeding by using $X(\gamma) = \varepsilon e^{\bar{b}\gamma}$ for $\gamma \geq 0$.

FIGURE 3

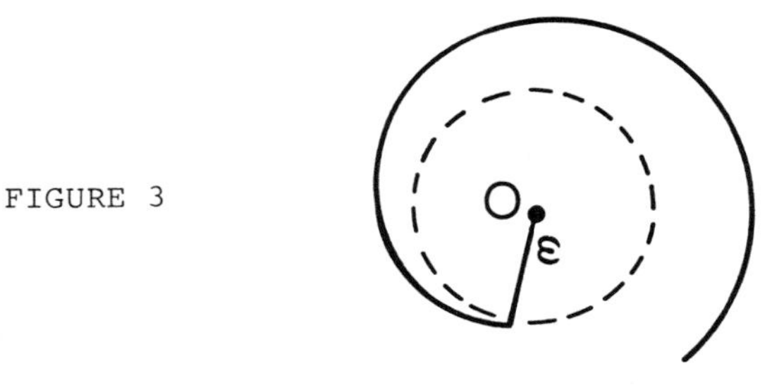


Remark 2. The search game just described takes place in the plane. However, the object sought is not a point but an infinite ray. (This ray is represented, in polar coordinates, by $(\alpha|H|, \beta)$ $\alpha \geq 1$.) Thus, the preceding problem is essentially one dimensional. This explains the fact that the capture time is normalized by $|H|$ (see (42)) and not by $|H|^2$.

## 8.3 "SWIMMING IN A FOG" PROBLEMS

The type of problem to be described in this section was presented by Bellman (1956) as a research problem. A narrative description of this problem is the following. A person has been shipwrecked at a point O in a fog and wishes to minimize the maximum time required to reach the shore, given its shape and some information about its location around O. Similar problems can be considered in which the distribution of the location of the boundary (shore) line is given and one would like to minimize the expected time to reach it.

Gross (1955) considered the following formulation of that problem. Find a shortest plane trajectory with the property that if the origin of the trajectory is covered in any way by a given plane figure, some point of the trajectory lies on the boundary of the figure. He presented a discussion about the nature of the solution for the cases in which this figure is

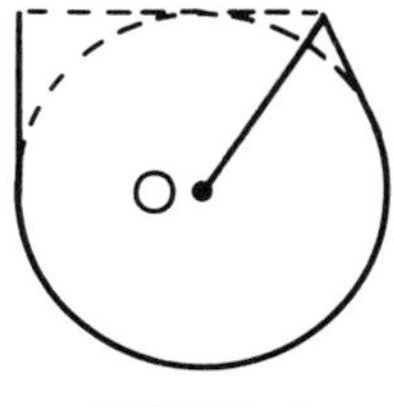


FIGURE 4

the circle, the equilateral triangle, a "keyhole"-shaped figure, and the infinite strip of unit width.

Isbell (1957) found the trajectory which guarantees reaching an infinite line with unit distance from O in minimum time, and briefly considered a two-line problem. The minimax search trajectory for a line of a unit distance as described by Isbell (1957) is the following. Being at O, imagine a clock face (Figure 4). Walk toward 1 o'clock for $\sqrt{4/3}$ units. Then turn on the tangent which strikes the unit circle at 2 o'clock. Follow the circle to 9 o'clock and continue on a tangent. Upon striking the line which is tangent to the unit circle at 12 o'clock, all the tangents to the unit circle have been swept. The maximum distance traveled in this minimax trajectory is $7\pi/6 + \sqrt{3} + 1 \sim 6.397$ units.

Gluss (1961a,b) presented a solution of the minimax search for a circle with a known radius, of known distance from O, and an approximate solution for minimizing the expected distance traveled in the case that the shore is a line of unit distance from O uniformly distributed around O.

The minimax search for a line can be considered within the framework described in Chapter 5.1. We obtain the following research problem. Let $|H|$ be the distance of the line from the origin. Given the information that the expectation of $|H|$ satisfies $E|H| \leq 1$, what are the minimax search trajectory (pure strategy) and the optimal search strategy (mixed) of the searcher?

In contrast to the solution obtained by Isbell (1957), we expect a smooth minimax searching trajectory. If we could show that a result similar to Theorem 5 of Section 6.4 holds for this problem, then the minimax trajectory would be an exponential spiral, and one could simply follow the technique presented in Section 8.2. However, the functional involved in this case does not seem to satisfy the unimodality condition. Thus, an extension of those results is needed.

# Appendixes

Appendix 1

# On The Existence Of A Value For Search Games

In this appendix we use some general theorems in functional analysis and game theory in order to show that search games with mobile or immobile hider have a value. We will assume that the search space Q is a connected subset of a Euclidean space. The search space Q will be required to be closed (i.e., to contain the set of its accumulation points), but will not be required to be bounded.

Using the framework described in Chapter 1, we let TS be the set of all (continuous) search trajectories with velocity not exceeding 1. Thus, any trajectory $S \in TS$ is a function $S(t)$ from the nonnegative numbers into Q, which, for any $0 \le t_1 < t_2$, satisfies $d(S(t_2), S(t_1)) \le t_2 - t_1$. We can either assume that $S(0)$ is the origin, or that $S(0)$ is any point chosen by the searcher from a certain closed bounded subset of Q.

We introduce the following topology on TS: $S_i \to S$ if $S_i(t) \to S(t)$ uniformly for any finite time interval. In this topology, a typical neighborhood of S is the set of trajectories that lie within a strip of width $2\varepsilon$ around S in an interval $0 \le t \le \bar{t}$. Under this topology, it follows from the foregoing assumptions that TS is closed, and it is easy to verify that the other requirements of the Ascoli theorem (see Kelley, 1955, p. 233) are also satisfied so that TS is compact (i.e., each family of open sets which covers TS has a finite subfamily which covers TS). TS is obviously a Hausdorff space (i.e., two distinct elements of TS have disjoint neighborhoods). Thus TS is a compact Hausdorff space.

A similar topology can be introduced on the set of (continuous) hiding trajectories TH. In this topology, the set TH is compact only if the search space is bounded and the velocity of the hider is bounded, but we shall not need the compactness of TH, so that our results are also valid in those cases for which either the search space or the velocity of the hider (or both) are unbounded.

We now consider some properties of the capture time C(S, H) defined as

$$C(S, H) = \begin{cases} \min\{t : d(S(t), H(t)) \leq r\} \\ \text{or, infinity if no such } t \text{ exists.} \end{cases} \tag{1}$$

It is easy to see that C(S, H) is not continuous in either of its variables, but it does have the fundamental property of lower semicontinuity defined as follows.

DEFINITION 1. A function f is called lower semicontinuous if $X_i \to X$ implies $\underline{\lim}\, f(X_i) \geq f(X)$.

An equivalent definition is that for any real $\alpha$, $\{X : f(X) > \alpha\}$ is open.

First we show that:

LEMMA 1. For any fixed H, C(S, H) is lower semicontinuous in S.

Proof. Assume that $S_i \to S$ and let $\alpha$ be any real number. If $C(S, H) > \alpha$, then $\min_{0 \le t \le \alpha} d(S(t), H(t)) \ge r + \delta$ for some $\delta > 0$. By the assumption, $S_i(t) \to S(t)$ uniformly for $0 \le t \le \alpha$, which implies that there exists an integer $i_0$ such that, for all $i > i_0$, $\inf_{0 \le t \le \alpha} d(S_i(t), H(t)) \ge r + \delta/2$. Thus $\underline{\lim}\, C(S_i, H) > \alpha$. Q.E.D.

An identical proof can be used to show:

LEMMA 2. For any fixed S, C(S, H) is lower semicontinuous in H.

The fact that C(S, H) is lower semicontinuous enables us to consider the payoff in the case that the players use mixed strategies defined as follows.

DEFINITION 2. A mixed strategy s (resp. h) of the searcher (resp. hider) is a regular Borel probability measure on TS (resp. TH). (A positive Borel measure M on TS is said to be regular if

$$M(E) = \sup\{M(A) : A \subset E\} = \inf\{M(U) : U \supset E\}$$

for every Borel set $E \subset TS$, where A ranges over the compact subsets of E and U ranges over the open sets which contain E (see Rudin, 1973).)

The set of all mixed strategies will be denoted by Ts (resp. Th).

Obviously, Ts and Th are convex sets. It follows from Lemmas 1 and 2 that C(S, H) is Borel measurable in both variables. We can thus define the payoff c(s, h) as the expected value of C in the cross product measure $s \times h$:

$$c(s, h) = \int C(S, H)\, d(s \times h). \qquad (2)$$

If the measure of the set $\{(S, H) : C(S, H) = \infty\}$ is positive under $s \times h$, then we define $c(s, h) = \infty$, while if the measure of that set is zero, then the integral appearing in (2) has its usual meaning.

Since $C(S, H)$ is nonnegative, it follows from Fubini's theorem that

$$c(s, h) = \iint C(S, H)\, ds\, dh = \iint C(S, H)\, dh\, ds. \tag{3}$$

We will show that the search game has a value, i.e., that

$$\inf_{s \in Ts} \sup_{h \in Th} c(s, h) = \sup_{h \in Th} \inf_{s \in Ts} c(s, h) \tag{4}$$

using the following theorem (see Fan, 1953):

FAN'S MINIMAX THEOREM. Let Ts be a compact Hausdorff space, Th any set, and $c(s, h)$ a function on $Ts \times Th$ which is convex in s and concave in h. If $c(s, h)$ is lower semicontinuous in s for each h, then (4) holds.

In order to use Fan's minimax theorem, we have to introduce a topology on Ts and show that this set is compact and Hausdorff. We also have to show that under this topology, $c(s, h)$ is a lower semicontinuous function of s. (All the other requirements are obviously satisfied since Ts and Th are convex sets and $c(s, h)$ is linear in s and in h.)

Let B(TS) be the Banach space of all real continuous functions on TS with the supremum norm. The Riesz representation theorem identifies the dual space $B(TS)^*$ with the space of all real Borel measures on TS that are differences of regular positive ones. The set Ts of all regular Borel probability measures on TS is a subset of the unit sphere in $B(TS)^*$.

The topology we introduce on Ts is the weak$^*$ topology (denoted by $w^*$). Under this topology, $s_i \overset{w^*}{\rightarrow} s$ iff $\int f(S)\, ds_i \rightarrow \int f(S)\, ds$ for all real continuous functions f defined on TS. It is well known (see, e.g., Rudin, 1973, p. 77) that the set of all regular Borel probability measures on a compact set is compact in the $w^*$ topology. Obviously, Ts is also Hausdorff.

We now make the last step needed to use the minimax theorem, by proving the following result.

LEMMA 3. For each h, c(s, h) is a lower semicontinuous function of s in the $w^*$ topology.

Proof. We shall use Fatou's lemma which states that if $f_i$ is any sequence of nonnegative integrable functions and M is a positive measure, then

$$\underline{\lim} \int f_i \, dM \geq \int \underline{\lim}\, f_i \, dM. \tag{5}$$

At first we show that for each h, c(S, h) is a lower semicontinuous function of S. Let $S_i \rightarrow S$. Then by Fatou's lemma and the lower semicontinuity of C, we obtain

$$\underline{\lim}\, c(S_i, h) = \underline{\lim} \int C(S_i, H)\, dh$$

$$\geq \int \underline{\lim}\, C(S_i, H)\, dh \geq \int C(S, H)\, dh = c(S, h).$$

We now show that c(s, h) is a lower semicontinuous function of s. Let h $\in$ Th be any fixed element and assume that $s_i \overset{w^*}{\rightarrow} s$. For any $t \geq 0$, consider the set $U(t) = \{S : C(S, h) > t\}$. Since C(S, h) is lower semicontinuous, it follows that U(t) is open. Let us define $P_i(t)$ as the probability of U(t) under $s_i$ and P(t) as the probability of U(t) under s. Since

U(t) is open, it follows that $\underline{\lim}_{i \to \infty} P_i(t) \geq P(t)$. This is a consequence of the following facts:

(1) If U is an open set in a compact Hausdorff space TS, then for any closed set $A \subset U$ there exists a continuous function f on TS to the interval [0, 1] such that f is one on A and zero on the complement of U (see Kelley, 1955, pp. 115 and 141).

(2) s is a regular measure.

(3) $\int f(S)\, ds_i \to \int f(S)\, ds$ for all real continuous functions.

Now,

$$\underline{\lim}\, c(s_i, h) = \underline{\lim} \int_0^\infty P_i(t)\, dt \geq \int_0^\infty \underline{\lim}\, P_i(t)\, dt$$

$$\geq \int_0^\infty P(t)\, dt = c(s, h). \qquad \text{Q.E.D.}$$

It follows from the preceding discussion that the conditions of Fan's minimax theorem are satisfied. Thus, we deduce that any search game in a compact space has a value. Moreover, since $v(s) = \sup_{h \in Th} c(s, h)$ is also a lower semicontinuous function of s and since Ts is compact, it follows that there exists an $s^* \in Ts$ which satisfies $v(s^*) = \inf_{s \in Ts} v(s)$. Thus, there exists an optimal strategy for the searcher.

Note that the hider may have only $\varepsilon$-optimal strategies, as happens, for example, in the search for a mobile hider on a circle with a fixed starting point.

If the search space is not bounded, then it follows from the preceding discussion that the search game has a value in the case that the capture time is used as a cost function and the hider's strategies are restricted to a convex subset of Th.

Since the restriction introduced in Chapter 5, $E|H|^N \leq \lambda$ ($|H|$ is the distance of the immobile hider from the origin, N the dimension of Q, and $\lambda$ a constant), confines h to a convex subset, it follows that any search game with such a structure has a value. In the case that we use a normalized cost function $\tilde{C}(S, H) = C(S, H)/|H|^N$, then it is easy to verify that $\tilde{C}(S, H)$ is lower semicontinuous in both variables, so that the same considerations used above show that such a game has a value and an optimal search strategy.

Remark 1. The results established in this appendix remain valid in the case that the maximal velocity of the searcher depends on his location and the detection radius depends on the location of the hider. In this case, the capture time C(S, H) is given by

$$C(S, H) = \begin{cases} \min\{t : d(S(t), H(t)) \leq r(H(t))\} \\ \text{or, infinity if no such } t \text{ exists,} \end{cases} \tag{6}$$

where r(Z) is a positive continuous function.

It can easily be established that C(S, H) given by (6) is a lower semicontinuous function both in S and in H. The proof is almost identical to the proof of Lemma 1. The only properties needed to prove the existence of a value and of an optimal search strategy are the lower semicontinuity of C(S, H) and the compactness of TS. Thus, it follows that these results also hold for nonhomogeneous search spaces.

Remark 2. Glicksberg (1950) proved the existence of a value for a game played over a pair of compact metric spaces, in which the payoff is a lower (upper) semicontinuous function. The result established in this appendix is stronger because TS is compact and Hausdorff but not a metric space, and TH is not even required to be compact.

Remark 3. Wilson (1977) proved the existence of a value for two-person differential games in which the players receive no information about the state variables during the game, except for their initial values, which are known to both players. However, Wilson's results depend on the following assumptions which do not hold in our case.

(1) The game is of prescribed duration.

(2) Each player chooses a control from a compact set.

(3) The payoff is an integral of a function which is continuous in the state variables.

Thus, even though Wilson's results are quite general, they are not applicable to search games.

Appendix 2

# Theorems About The Attainment Of Exponential Functions

In this appendix, we show that for any positive function (resp. sequence) $X(\theta)$ defined for $0 \leq \theta < \infty$, the cone spanned by the family $\{X^{+\gamma}, 0 \leq \gamma < \infty\}$, where $X^{+\gamma}$ is defined by

$$X^{+\gamma}(\theta) = X(\gamma + \theta), \tag{1}$$

contains a function (resp. a sequence) which is "close enough" to an exponential function (resp. a geometric sequence). The theorems proved in this appendix are used in establishing the results presented in Chapter 6.

We start with the discrete case.

THEOREM 1. Let $x_i$, $0 \leqq i < \infty$, be a positive sequence and let $k > 0$ be any integer. Let $W_k$ be the $(k + 1)$-dimensional convex cone spanned by the set $\{(x_i, x_{i+1}, \ldots, x_{i+k}), 0 \leqq i < \infty\}$. Thus,

$$W_k = \left\{ Y : Y = \sum_{i=0}^{n} \beta_i \cdot (x_i, x_{i+1}, \ldots, x_{i+k}); \right.$$
$$\left. \beta_i \geqq 0,\ 0 \leqq i \leqq n,\ 0 < n < \infty \right\}. \tag{2}$$

$\overline{W}_k$ will denote the closure of $W_k$. Define

$$a = \overline{\lim_{n\to\infty}}\, x_n^{1/n}. \tag{3}$$

Then

$$\frac{1}{\sum_{i=0}^{k} a^i} (1, a, a^2, \ldots, a^k) \in \overline{W}_k. \tag{4}$$

(The case where $a = \infty$ is included. In this case (4) means: $(0, 0, \ldots, 0, 1) \in \overline{W}_k$.)

Proof. We consider three different cases.

1. If

$$a = \overline{\lim_{n\to\infty}}\, x_n^{1/n} = \infty, \tag{5}$$

then we can show that

$$\overline{\lim_{n\to\infty}}\, \frac{x_{n+k}}{x_n + x_{n+1} + \cdots + x_{n+k-1}} = \infty. \tag{6}$$

The last statement follows from the consideration that if there exists a positive number $\alpha$ such that for any $n > 0$,

$$x_{n+k} < \alpha(x_n + x_{n+1} + \cdots + x_{n+k-1}), \tag{7}$$

then we could prove by induction that there exists a positive constant $\beta$ such that for all $n > 0$,

$$x_n < \beta(\alpha + 1)^n,$$

which obviously contradicts (5).

Thus if (5) holds, then (6) holds. Hence for any $\varepsilon > 0$, we can find an integer $n$ such that for all $0 \leqq j < k$,

$$x_{n+j}/x_{n+k} < \varepsilon. \tag{8}$$

It follows that the $(k + 1)$-tuple

$$(1/x_{n+k})(x_n, x_{n+1}, \ldots, x_{n+k})$$

is as close as desired to $(0, 0, \ldots, 0, 1)$.

2. In this case, we assume that

$$a = \overline{\lim_{n\to\infty}} x_n^{1/n} = 0. \tag{9}$$

We shall show that (9) implies

$$\overline{\lim_{n\to\infty}} \frac{x_n}{x_{n+1} + \ldots + x_{n+k}} = \infty. \tag{10}$$

This result is established as follows. If (10) does not hold, then there exists a positive number $\alpha$ such that for any $n > 0$,

$$x_n < \alpha(x_{n+1} + \ldots + x_{n+k}) < \alpha \sum_{n+1}^{\infty} x_i. \tag{11}$$

If we define

$$y_n = \sum_{n}^{\infty} x_i, \tag{12}$$

we may note that (9) implies that this sum is finite, and then (11) implies

$$y_n - y_{n+1} < \alpha y_{n+1}$$

or

$$y_{n+1} > \frac{1}{\alpha + 1} y_n > \ldots > \left(\frac{1}{\alpha + 1}\right)^n y_1, \tag{13}$$

which is easily shown to contradict (9).

Thus if (9) holds, then (10) holds, so that in this case we can find an integer n such that the $(k + 1)$-tuple $(1/x_n)(x_n, x_{n+1}, \ldots, x_{n+k})$ is as close as desired to $(1, 0, 0, \ldots, 0)$.

3. If (5) and (9) do not hold, then we can assume that a defined by (3) satisfies

$$0 < a < \infty. \tag{14}$$

We shall show that for any $\varepsilon > 0$, there exists a sequence $\beta_i \geqq 0$, $0 \leqq i < \infty$, so that the numbers $D_j$, $0 \leqq j \leqq k$, defined by

$$D_j = \sum_{i=0}^{\infty} \beta_i x_{i+j} \tag{15}$$

satisfy for all $0 \leqq j \leqq k - 1$,

$$\frac{a(1 - \varepsilon)}{1 + \varepsilon} < \frac{D_{j+1}}{D_j} < \frac{a}{1 - \varepsilon}. \tag{16}$$

The proof is based on the fact that the convolution of any sequence with a sequence which is "close enough" to a geometric sequence is also "close enough" to a geometric sequence, if it converges. We shall use the following construction.

Define a positive sequence $a_i$, $0 \leqq i < \infty$, satisfying

$$a_i \to a, \tag{17}$$

$$(1 - \varepsilon)a < a_i < a, \tag{18}$$

and

$$\sum_{i=0}^{\infty} \frac{x_{i+j}}{\Pi_{l=0}^{i} a_l} = \infty \qquad \text{for all} \quad 0 \leqq j \leqq k. \tag{19}$$

The existence of such a sequence follows from (3), which implies that for each $0 \leqq j \leqq k$, we can choose $a_l$ such that $\Pi_{l=0}^{i} a_l \leqq x_{i+j}$ for infinitely many i. Such a sequence can be defined by

$$a_i = (1 - (\varepsilon/l))a \qquad \text{for} \quad i_{l-1} < i \leqq i_l, \quad 1 \leqq l < \infty.$$

It follows from (3) that $i_l$ can be chosen large enough so that for each $0 \leqq j \leqq k$, there exists a positive integer

$m \leqq i_l - i_{l-1}$ satisfying

$$\left(\prod_{i=1}^{i_{l-1}} a_i\right)\left(1 - \frac{\varepsilon}{l}\right)^m a^m \leqq x_{i_{l-1}+m+j}.$$

The sequence $\beta_i$, $0 \leqq i < \infty$, is defined by

$$\beta_i = (1 - \delta)^i \Big/ \prod_{l=0}^{i} a_l, \tag{20}$$

where $\delta$ is a positive number satisfying

$$\delta \leqq \varepsilon \tag{21}$$

and

$$\beta_0 x_j \Big/ \sum_{i=1}^{\infty} \beta_i x_{i+j} < \varepsilon \qquad \text{for all} \quad 0 \leqq j \leqq k - 1. \tag{22}$$

One can show that the existence of a $\delta$ satisfying (20)-(22) follows from (19). It follows from (17) that for all $0 \leqq j \leqq k$,

$$\overline{\lim_{i\to\infty}}(\beta_i x_{i+j})^{1/i} = \overline{\lim_{i\to\infty}}\left(\frac{(1 - \delta)^i}{\prod_{l=0}^{i} a_l} x_{i+j}\right)^{1/i} = 1 - \delta < 1,$$

which implies that for all $0 \leqq j \leqq k$,

$$D_j = \sum_{i=0}^{\infty} \beta_i x_{i+j} < \infty. \tag{23}$$

We shall show that these $D_j$, $0 \leqq j \leqq k$, satisfy (16). To this end we denote

$$B_j = \sum_{i=0}^{\infty} \beta_i x_{i+j+1} \Big/ \sum_{i=0}^{\infty} \beta_{i+1} x_{i+j+1}. \tag{24}$$

It follows from (18), (20), and (21) that

$$B_j \leqq \sup_{0 \leqq i < \infty} \frac{\beta_i}{\beta_{i+1}} = \sup_{0 \leqq i < \infty} \frac{a_{i+1}}{1 - \delta} < \frac{a}{1 - \delta} < \frac{a}{1 - \varepsilon} \tag{25}$$

and

$$B_j \geqq \inf_{0 \leqq i < \infty} \frac{\beta_i}{\beta_{i+1}} = \inf_{0 \leqq i < \infty} \frac{a_{i+1}}{1 - \delta} > a(1 - \varepsilon). \tag{26}$$

Inequality (16) can now be established as follows:

$$\frac{D_{j+1}}{D_j} = B_j \frac{1}{\left(\beta_0 x_j \Big/ \left(\sum_{i=0}^{\infty} \beta_{i+1} x_{i+j+1}\right)\right) + 1}$$

(by (23) and (24)).

Using (22), (25), and (26), we obtain

$$\frac{a(1 - \varepsilon)}{1 + \varepsilon} < \frac{B_j}{1 + \varepsilon} < \frac{D_{j+1}}{D_j} < B_j < \frac{a}{1 - \varepsilon}, \tag{27}$$

so that (16) follows directly from (27).

The proof can be now completed by choosing an n such that for all $0 \leqq j \leqq k$, the sum $\sum_{i=0}^{n} \beta_i x_{i+j}$ will be close enough to $D_j$. Q.E.D.

Two remarks about Theorem 1 might be helpful.

Remark 1. In order for Theorem 1 to be valid, it is necessary to consider the closure of $W_k$ defined by (2). $W_k$ itself will not always contain a "geometric point" defined by (4), even if we allow infinite sums in the definition (2), as the following example illustrates.

Let $x_i = 1/i$, $1 \leqq i < \infty$, and let $k = 2$. Then for any nonnegative sequence $\beta_i$,

$$\sum_{i=1}^{\infty} \beta_i \left(\frac{1}{i}, \frac{1}{i + 1}, \frac{1}{i + 2}\right)$$

$$= \left(\sum_{i=1}^{\infty} \frac{\beta_i}{i}, \sum_{i=1}^{\infty} \frac{\beta_i}{i + 1}, \sum_{i=1}^{\infty} \frac{\beta_i}{i + 2}\right)$$

$$= (c_0, c_1, c_2). \tag{28}$$

It can be easily verified by definition (28) that $c_0c_2 > c_1^2$. (This can be done by comparing the coefficients of $\beta_i^2$: $1/(i(i + 2)) > 1/(i + 1)^2$ and the coefficients of $\beta_i\beta_j$ for $i \neq j$: $1/(i(j + 2)) + 1/(j(i + 2)) > 2/((i + 1)(j + 1))$.) Thus $W_2$ does not contain any "geometric point."

On the other hand,

$$\lim_{i\to\infty} i\left(\frac{1}{i}, \frac{1}{i + 1}, \frac{1}{i + 2}\right) = (1, 1, 1),$$

so that by using $\beta_i = i$ and $\beta_j = 0$ for $j \neq i$ ($i$ being a "large" number), we can achieve a 3-tuple as close as desired to a "geometric point."

Remark 2. The geometric point appearing in (4), where $a$ is defined by (3), is not necessarily unique. This can be illustrated by the following example. Let $d$ be a positive number different from 1 and define the sequence $X = \{x_i\}$ as

$$X = \left\{\frac{d^0}{d^0}, \frac{d^0}{d}, \frac{d}{d}, \frac{d^0}{d^2}, \frac{d}{d^2}, \frac{d^2}{d^2}, \frac{d^0}{d^3}, \frac{d}{d^3}, \frac{d^2}{d^3}, \frac{d^3}{d^3}, \ldots, \frac{d^0}{d^\ell}, \frac{d}{d^\ell}, \frac{d^2}{d^\ell}, \ldots, \frac{d^\ell}{d^\ell}, \ldots\right\}.$$

Then clearly,

$$a = \overline{\lim_{n\to\infty}}\, x_n^{1/n} = 1,$$

so that for any $k > 0$, $(1, 1, \ldots, 1)$ belongs to $W_k$.

On the other hand, it is obvious that $(1, d, \ldots, d^k)$ also belongs to $\overline{W}_k$. (Use $\beta_{k(k+1)/2} = d^k$ and $\beta_j = 0$ for $j \neq k(k + 1)/2$.)

Of course, there may be cases where the geometric point defined by (3) and (4) is the only one belonging to $\overline{W}_k$. A trivial case of this kind occurs when $X$ is itself a geometric sequence.

Continuous Theorem

We now present a continuous version of Theorem 1.

THEOREM 1′. Let $X(\theta)$, $0 \leqq \theta < \infty$, be any measurable positive function which is bounded on any finite interval. Define

$$b = \ln\left[\overline{\lim_{n\to\infty}}\left(\int_n^{n+1} X(\theta)\, d\theta\right)^{1/n}\right]. \tag{29}$$

Suppose $-\infty < b < \infty$. For any $\varepsilon > 0$, $\bar{\varepsilon} > 0$, and $D > 0$, there exists a positive number $L$ and a nonnegative function $\beta(t)$, $0 \leqq t < \infty$, such that $\beta'(t)$ is continuous and the function $f(\theta)$ defined by

$$f(\theta) = \int_0^L \beta(t)X(t + \theta)\, dt$$

satisfies

$$(1 - \varepsilon)e^{b\theta} \leqq f(\theta) \leqq (1 + \varepsilon)e^{b\theta}, \tag{30}$$

and

$$(1 - \bar{\varepsilon})be^{b\theta} \leqq f'(\theta) \leqq (1 + \bar{\varepsilon})be^{b\theta} \tag{30′}$$

for all $0 \leqq \theta \leqq D$.

Proof. The proof is similar to the one presented for Theorem 1. We shall present the general lines of the proof, while the missing details can be established by drawing analogies from the proof of Theorem 1.

Let $\varepsilon_1 > 0$ be a small positive number. As in the proof of Theorem 1, we define a continuous and monotone increasing function $b(u)$, $0 \leqq u < \infty$, which satisfies

$$\left.\begin{aligned} &\lim_{u\to\infty} b(u) = b && \text{(see (29))}, \\ &(1 - \varepsilon_1)b \leqq b(u) < b && \text{for all } 0 \leqq u < \infty, \end{aligned}\right\} \tag{31}$$

and

$$\int_0^\infty X(t+\theta) \exp\left[-\int_0^t b(u)\, du\right] dt = \infty \tag{31'}$$

$$\text{for all } 0 \leqq \theta \leqq D+1.$$

Equation (31´) can be established as follows. For all $0 \leqq \theta \leqq D + 1$,

$$\int_0^\infty X(t+\theta) \exp\left[-\int_0^t b(u)\, du\right] dt$$

$$\geqq \int_0^\infty X(t+\theta) \exp\left[-\int_\theta^{t+\theta} b(u)\, du\right] dt$$

(since $b(u)$ is monotone)

$$= \exp\left[\int_0^\theta b(u)\, du\right] \cdot \int_0^\infty X(t+\theta) \exp\left[-\int_0^{t+\theta} b(u)\, du\right] dt$$

$$= \exp\left[\int_0^\theta b(u)\, du\right] \cdot \int_\theta^\infty X(t) \exp\left[-\int_0^t b(u)\, du\right] dt.$$

Thus, it is sufficient to choose $b(u)$ so as to satisfy

$$\int_0^\infty X(t) \exp\left[-\int_0^t b(u)\, du\right] dt = \infty.$$

This can be done just as in Theorem 1, taking care to make $b(u)$ continuous.

We define a function $\beta_1(t)$, $0 \leqq t < \infty$, by

$$\beta_1(t) = \alpha_1 \exp\left[-\delta t - \int_0^t b(u)\, du\right], \tag{32}$$

where $\alpha_1$ is a positive constant to be determined later and $\delta$ satisfies

$$0 < \delta < \varepsilon_1$$

and

$$\int_0^{D+1} X(t)\beta_1(t)\, dt \bigg/ \int_{D+1}^\infty X(t)\beta_1(t)\, dt < \varepsilon_1.$$

Now, it follows from (29), (31), and (32) that the integral

$$f_1(\theta) = \int_0^\infty \beta_1(t)X(t + \theta)\ dt \tag{33}$$

exists and is finite and convergence is uniform for $0 \leqq \theta \leqq D + 1$.

Using a method similar to the one used in the proof of Theorem 1, it can be easily verified that for all $0 \leqq \theta \leqq D + 1$,

$$(1 - \varepsilon)e^{b\theta} < f_1(\theta)/f_1(0) < (1 + \varepsilon)e^{b\theta}.$$

For any arbitrary positive constant $\alpha$, we can choose $\alpha_1$ in (32) so that $f_1(0) = \alpha$. Thus

$$\alpha(1 - \varepsilon)e^{b\theta} < f_1(\theta) < \alpha(1 + \varepsilon)e^{b\theta}.$$

On the other hand, it follows from (33) that there exists a positive number L so that $f_2(\theta)$ defined by

$$f_2(\theta) = \int_0^{L-1} \beta_1(t)X(t + \theta)\ dt$$

satisfies for all $0 \leqq \theta \leqq D + 1$

$$\alpha(1 - \varepsilon)e^{b\theta} \leqq f_2(\theta) \leqq \alpha(1 + \varepsilon)e^{b\theta}. \tag{34}$$

Now, define $f(\theta)$ as

$$f(\theta) = \int_\theta^{\theta+1} f_2(t)\ dt = \int_0^L \beta(t)X(t + \theta)\ dt, \tag{35}$$

where

$$\begin{aligned} \beta(t) &= \int_0^t \beta_1(t - u)\ du && \text{for} \quad 0 \leqq t \leqq 1, \\ &= \int_0^1 \beta_1(t - u)\ du && \text{for} \quad 1 < t \leqq L - 1, \\ &= \int_{t-L-1}^1 \beta_1(t - u)\ du && \text{for} \quad L - 1 < t \leqq L. \end{aligned}$$

It follows from (34) and (35) that

$$(1 - \varepsilon)\ \frac{\alpha(e^{b} - 1)}{b}\ e^{b\theta} \leqq f(\theta) \leqq \alpha(1 + \varepsilon) \int_{\theta}^{\theta+1} e^{bt}\, dt$$

$$= (1 + \varepsilon)\ \frac{\alpha(e^{b} - 1)}{b}\ e^{b\theta}.$$

If we choose

$$\alpha = b/(e^{b} - 1), \tag{36}$$

then (30) holds.

In order to prove (30′), we note that

$$f'(\theta) = f_2(\theta + 1) - f_2(\theta).$$

Thus, it follows from (34) that

$$\alpha(1 - \varepsilon)e^{b(\theta+1)} - \alpha(1 + \varepsilon)e^{b\theta}$$

$$\leqq f'(\theta) \leqq \alpha(1 + \varepsilon)e^{b(\theta+1)} - \alpha(1 - \varepsilon)e^{b\theta}.$$

Using (36), we obtain

$$be^{b\theta} - \varepsilon R_1(\theta) \leqq f'(\theta) \leqq be^{b\theta} + \varepsilon R_2(\theta),$$

where $R_1(\theta)$ and $R_2(\theta)$ are bounded for $0 \leqq \theta \leqq D$. This proves (30′). Q.E.D.

Appendix 3

# Discrete Search Games

In this appendix, we briefly describe some discrete search games which appear in the literature but do not fall into the general framework described in the text.

In the games to be described, the search space Q consists of n boxes, i.e., $Q = \{1, \ldots, n\}$. A pure strategy S of the searcher is a mapping from the natural numbers into Q, where $S(i) = K$ means that in his i-th move the searcher looks in the K-th box. A pure strategy of the hider is a number $K \in Q$ in the search games with immobile hider, or a mapping from the natural numbers into Q in the case that the hider is mobile. There is a conditional probability $P_K$ of detection if the searcher looks in the K-th box and the hider is there. There is a cost $R_K$ for looking in the i-th box ($R_K$ is identically one for some of the games) and the cost function is the total sum paid by the searcher.

Bram (1963) considers the case of an immobile hider with $R_K = 1$ and presents some properties of this game. Neuts (1963) solves this game with general $R_K$ under the assumption that the searcher is restricted to the use of stationary strategies, which is equivalent to assuming that the searcher is memoryless.

Norris (1962) solves some games with a hider who sees where the searcher has looked, and can move from one box to another between looks. A cost is usually associated with such a move. In the special case that this cost of moving is zero, then the minimax hiding strategy consists of hiding initially and after each successive search, using the location probability vector $U^* = (u_1^*, u_2^*, \ldots, u_n^*)$ defined such that $u_K^* P_K / R_K$ is the same for all K. Roberts and Gittins (1978) investigate the goodness of the hiding strategy $U^*$ as an approximation to the optimal hiding strategy for the case that the hider is immobile and $R_K = 1$.

Gilbert (1962) and Johnson (1964) consider the discrete search game, with immobile hider, with $R_K = 1$ under the additional assumption that if the searcher looks in the K-th box and does not detect the hider in that box, he still receives the information as to whether the number of the box he has just searched is greater or less than the number of the box which contains the hider. Solutions of this game were given only for $n \leq 11$. Gal (1974b) considers a discrete search game in which the immobile hider chooses a number $H \in Q$ and at each step, the searcher gains the binary information $H \leq S(i)$ or $H > S(i)$. The game proceeds until the searcher locates H, with the cost being the number of steps made by the searcher during the game. A general solution for this game is presented. The author shows that the optimal strategies of both players are generally mixed strategies which are not unique and that the natural bisection strategy of the searcher is usually not optimal. A continuous problem of this type in which H and S(i) are real numbers in a certain interval and the information $H \leq S(i)$ or $H > S(i)$ has a certain probability of being erroneous is presented and solved by Gal (1978). Another continuous version in which the searcher receives the binary information $H \leq S(i)$ or $H > S(i)$ and there is a travel cost of $b|S(i + 1) - S(i)|$, where b is a constant, in addition to the fixed cost of observation, is solved by Murakami (1976).

Appendix 4

# Some Basic Notions Of The Theory Of Games

The mathematical theory of games was developed by Von Neumann and published for the first time in 1943 (Von Neumann and Morgenstern, 1953). There are several other books which can serve as a good introduction to the theory of games, but we shall refer to only one of them: "Games and Decisions" (Luce and Raiffa, 1957).

A game is essentially a set of rules describing the formal structure of a competitive situation. Any particular instance of a game is called a "play." The rules of the game, which are known to the players, specify the set of pure strategies available to each one of the players. A pure strategy is a plan formulated by a player prior to a play, which will cover all the possible decisions which may confront him during any play of the game. These strategies are called "pure strategies" in order to distinguish between them and "mixed strategies" which are probabilistic choices from the set of pure strategies. The expected course of a play is determined by the selection of a pure strategy by each player in ignorance of that chosen by any other player. The "payoffs" of a game are summarized as functions of the different (pure) strategies for each player. A game with two players where any gain of one player equals a corresponding loss of the other player is known as a "two-person zero-sum game." In such a game, it suffices to express the outcomes in terms of the payoff to one player. (In the main text, we shall usually refer to this payoff as the "cost function.") In this appendix, we shall deal only with two-person zero-sum games.

The approach which is used for solving two-person zero-sum games is to apply the minimax criterion: To accommodate the fact that each opponent is working against the other's interest, the minimax criterion selects for each player a strategy which yields the best of the worst possible outcomes. An "optimal solution" is said to be reached if neither player finds it beneficial to alter his strategy. In this case, the game is said to be in a state of equilibrium.

Keeping this framework in mind, we first consider games in which the number of pure strategies is finite. In this case, the game can be described by an m by n matrix $C = (C_{ij})$, where each entry $C_{ij}$ represents the amount that the first player (in the main text the hider) receives from the second player (in the main text the searcher) if the first player uses his i-th pure strategy and the second player uses his j-th pure strategy. It may happen that the matrix C will have a saddle point, i.e., an element $C_{i^*j^*}$ such that

$$\max_{1 \le i \le m} C_{ij^*} = C_{i^*j^*} = \min_{1 \le j \le n} C_{i^*j} .$$

In this case, the game would be in a state of equilibrium if the first player chooses his $i^*$-th pure strategy and the second player chooses his $j^*$-th pure strategy. The preceding strategies would be optimal, and thus this game could be solved using only pure strategies. Usually, however, such a saddle point does not exist because even in the simplest games (e.g., matching pennies, where $C = \begin{pmatrix} -1 & +1 \\ +1 & -1 \end{pmatrix}$), a player is at a disadvantage if he always uses the same pure strategy. The fact that usually there exists no optimal strategy in the set of pure strategies has led to the idea of using mixed strategies. Each player, instead of selecting a specific pure strategy, may choose an element from the set of his pure strategies according to a predetermined set of probabilities. Mixed strategies for the first and the second players will be denoted by $X = (x_1, \ldots, x_m)$ and $Y = (y_1, \ldots, y_n)$, respectively, where $x_i$, $i = 1, \ldots, m$, is the probability that the first player will choose his i-th pure strategy and $y_j$, $j = 1, \ldots, n$, is the probability that the second player will choose his j-th pure strategy.

When the first player plays a mixed strategy X and the second player a mixed strategy Y, the expected payment is given by the function

$$c(X, Y) = \sum_i \sum_j c_{ij} x_i y_j.$$

The fundamental theorem of two-person zero-sum finite games states that

$$\max_X \min_Y c(X, Y) = \min_Y \max_X c(X, Y).$$

This minimax value of c is called the "value" of the game and denoted by v. An equivalent result is that there exists a pair of mixed strategies $X^*$ and $Y^*$ such that

$$\max_X c(X, Y^*) = c(X^*, Y^*) \quad (= v) \quad = \min_Y c(X^*, Y).$$

Thus, c has a saddle point, or in other words, if the first player chooses the mixed strategy $X^*$ and the second player uses the mixed strategy $Y^*$, then each of them can guarantee an expected payoff of v. Thus, $(X^*, Y^*)$ is a pair of optimal strategies and the game has a solution in mixed strategies. (For example, in the matching pennies game, $X^* = (1/2, 1/2)$, $Y^* = (1/2, 1/2)$, and $v = 0$.)

The situation is more complicated if the game has an infinite number of pure strategies. In this case, the mixed strategies are probability measures on the set of pure strategies. Such a game is said to have a value v if, for any positive $\varepsilon$, the first player has a (mixed) strategy $X_\varepsilon$ which guarantees him an expected payoff of at least $(1 - \varepsilon)v$ and the second player has a (mixed) strategy $Y_\varepsilon$ which limits him to an expected loss of at most $(1 + \varepsilon)v$. If one of the players has an infinite number of pure strategies while the other player has only a finite number of pure strategies, then the game has a value. However, if both players have an infinite number of pure strategies, then the existence of a value is not assured. (For details see Luce and Raiffa, 1957, Appendix 7.)

In the _search_ games considered in the main text, both players have an infinite number of pure strategies. Nevertheless, we prove in Appendix 1 that any search game has a value.

# References


Agin, N. I. (1967). The application of game theory to ASW detection problems, Math. Rep., Princeton, New Jersey (September).

Alpern, S. (1974). The search game with mobile hider on the circle. In "Differential Games and Control Theory" (E. O Roxin, P. T. Liu, and R. L. Sternberg, eds.), pp. 181-200 Dekker, New York.

Arnold, R. D. (1962). Avoidance in one dimension: A continuous-matrix game. Operations Evaluation Group, Office of Chief of Naval Operations, Washington, D. C., OEG IRM-10 (AD 277 843), 14 pp. (January).

Beck, A. (1964). On the linear search problem. Israel J. Math. 2, 221-228.

Beck, A. (1965). More on the linear search problem. Israel J. Math. 3, 61-70.

Beck, A., and Newman, D. J. (1970). Yet more on the linear search problem. Israel J. Math. 8, 419-429.

Beck, A., and Warren, P. (1973). The return of the linear search problem. Israel J. Math. 14, 169-183.

Bellman, R. (1956). Minimization problem. Bull. Amer. Math. Soc. 62, 270.

Bellman, R. (1963). An optimal search problem. SIAM Rev. 5, 274.

Beltrami, E. J. (1963). The density of coverage by random patrols. NATO Conf. Appl. Oper. Res. to the Search and Detection of Submarines (J. M. Dobbie and G. R. Lindsey, eds.) 1, 131-148 (November).

Berge, C. (1973). "Graphs and Hypergraphs." North-Holland Publ., Amsterdam.

Bram, J. (1963). A 2-player N-region search game. Operations Evaluation Group, Office of Chief of Naval Operations, Washington, D. C., OEG IRM-31 (AD 402 914), 21 pp. (January).

Breiman, L. (1968). "Probability." Addison-Wesley, Reading, Massachusetts.

Christofides, N. (1975). "Graph Theory: An Algorithmic Approach." Academic Press, New York.

Danskin, J. M. (1968). A helicopter versus submarine search game. Oper. Res. 16, 509-517.

Dobbie, J. M. (1968). A survey of search theory. Oper. Res. 16, 525-537.

Edmonds, J. (1965). The Chinese postman problem. Bull. Oper. Res. Soc. Amer. 13, Suppl. 1, B-73.

Edmonds, J., and Johnson, E. L. (1973). Matching Euler tours and the Chinese postman problem. Math. Programming 5, 88-124.

Fan, K. (1953). Minimax theorems. Proc. Nat. Acad. Sci. U. S. A. 39, 42-47.

Feller, W. (1971). "An Introduction to Probability Theory and Its Applications," Vol. II, 2nd ed. Wiley, New York.

Foreman, J. G. (1974). The princess and the monster on the circle. In "Differential Games and Control Theory" (E. O. Roxin, P. T. Liu, and R. L. Sternberg, eds.), pp. 231-240. Dekker, New York.

Foreman, J. G. (1977). Differential search game with mobile hider. SIAM J. Control Optim. 15, 841-856.

Franck, W. (1965). On an optimal search problem. SIAM Rev. 7, 503-512.

Fristedt, B., and Heath, D. (1974). Searching for a particle on the real line. Adv. in Appl. Probab. 6, 79-102.

Gal, S. (1972). A general search game. Israel J. Math. 12, 32-45.

Gal, S. (1974a). Minimax solutions for linear search problems. SIAM J. Appl. Math. 27, 17-30.

Gal, S. (1974b). A discrete search game. SIAM J. Appl. Math. 27, 641-648.

Gal, S. (1978). A stochastic search game. SIAM J. Appl. Math. 31, 205-210.

Gal, S. (1979). Search games with mobile and immobile hider. SIAM J. Control Optim. 17, 99-122.

Gal, S., and Chazan, D. (1976). On the optimality of the exponential functions for some minimax problems. SIAM J. Appl. Math. 30, 324-348.

Gilbert, E. (1962). Games of identification and convergence. SIAM Rev. 4, 16-24.

Glicksberg, I. L. (1950). Minimax theorem for upper and lower semi-continuous payoffs. Rand Corp., Res. Memo. RM-478, 4 pp. (October).

Gluss, B. (1961a). An alternative solution to the "lost at sea" problem. Naval Res. Logist. Quart. 8, 117-121.

Gluss, B. (1961b). The minimax path in a search for a circle in the plane. Naval Res. Logist. Quart. 8, 357-360.

Gross, O. (1955). A search problem due to Bellman. Rand Corp., RM-1603 (AD 87 962), 8 pp. (September).

Gross, O. (1964). The rendezvous value of a metric space. In "Advances in Game Theory" (M. Dresher, L. S. Shapley, and A. Tucker, eds.), pp. 49-53. Princeton Univ. Press, Princeton, New Jersey.

Hardy, G. H., Littlewood, J. E., and Pólya, G. (1952). "Inequalities." Cambridge Univ. Press, London and New York.

Houdebine, Asp. J. (1963). Etude sur l'efficacité d'un barrage anti-sous-marin constitué. NATO Conf. Appl. Oper. Res. to the Search and Detection of Submarines (J. M. Dobbie and G. R. Lindsey, eds.) 1, 103-120 (November).

Isaacs, R. (1965). "Differential Games." Wiley, New York.

Isbell, J. R. (1957). An optimal search pattern. Naval Res. Logist. Quart. 4, 357-359.

Johnson, S. M. (1964). A search game. In "Advances in Game Theory" (M. Dresher, L. S. Shaply, and A. W. Tucker, eds.), pp. 39-48. Princeton Univ. Press, Princeton, New Jersey.

Kelley, J. L. (1955). "General Topology." Van Nostrand Reinhold, Princeton, New Jersey.

Koopman, B. O. (1946). Search and screening. Operations Evaluation Group Rep. No. 56, Center for Naval Analysis, Rosslyn, Virginia.

Koopman, B. O. (1956). The theory of search, Part II. Target detection. Oper. Res. 4, 503-531.

Langford, E. S. (1973). A continuous submarine versus submarine game. Naval Res. Logist. Quart. 20, 405-417.

Lawler, E. L. (1976). "Combinatorial Optimization: Networks and Matroids." Holt, New York.

Lindsey, G. R. (1968). Interception strategy based on intermittent information. Oper. Res. 16, 489-508.

Luce, R. D., and Raiffa, H. (1957). "Games and Decisions." Wiley, New York.

McCabe, B. J. (1974). Searching for a one-dimensional random walker. J. Appl. Probab. 11, 86-93.

Megiddo, N., and Hakimi, S. L. (1978). Pursuing mobile hider in a graph. The Center for Math. Studies in Econom. and Management Sci., Northwestern Univ., Evanston, Illinois, Disc. paper No. 360, 25 pp. (December).

Murakami, S. (1976). A dichotomous search with travel cost. J. Oper. Res. Soc. Japan 19, 245-254.

Neuts, M. F. (1963). A multistage search game. J. SIAM 11, 502-507.

Norris, R. C. (1962). Studies in search for a conscious evader. Lincoln Lab., MIT Tech. Rep. No. 279 (AD 294 832), 134 pp. (September).

Parsons, T. D. (1978a). Pursuit-evasion in a graph. In "Theory and Application of Graphs" (Y. Alavi and P. R. Lick, eds.). Springer-Verlag, Berlin.

Parsons, T. D. (1978b). The search number of a connected graph. Proc. Southwestern Conf. Combinatorics, Graph Theory, and Computing, 9th, Boca Raton, Florida (January-February).

Pavillon, Capt. de C. (1963). Un probleme de recherche sur zone. NATO Conf. Appl. Oper. Res. to the Search and Detection of Submarines (J. M. Dobbie and G. R. Lindsey, eds.) 1, 90-102 (November).

Roberts, D. M., and Gittins, J. C. (1978). The search for an intelligent evader: Strategies for searcher and evader in the two region problem. Naval Res. Logist. Quart. 25, 95-106.

Rudin, W. (1973). "Functional Analysis," McGraw-Hill, New York.

Stone, L. D. (1975). "Theory of Optimal Search." Academic Press, New York.

Von Neumann, J., and Morgenstern, O. (1953). "Theory of Games and Economic Behaviour." Princeton Univ. Press, Princeton, New Jersey.

Wilson, D. J. (1972). Isaacs' princess and monster game on the circle. J. Optim. Theory Appl. 9, 265-288.

Wilson, D. J. (1977). Differential games with no information. SIAM J. Control Optim. 15, 233-246.

Worsham, R. H. (1974). A discrete game with a mobile hider. In "Differential Games and Control Theory" (E. O. Roxin, P. T. Liu, and R. L. Sternberg, eds.), pp. 201-230. Dekker, New York.

Zelikin, M. I. (1972). On a differential game with incomplete information. Soviet Math. Dokl. 13, 228-231.

# Index